Criminal Justice
Recent Scholarship

Edited by
Nicholas P. Lovrich

A Series from LFB Scholarly

Estimating Drug Use
The Effect of Sampling Methods and Policy Implications

Janine Kremling

LFB Scholarly Publishing LLC
El Paso 2013

Library of Congress Cataloging-in-Publication Data

Kremling, Janine, 1977-
Estimating drug use : the effect of sampling methods and policy implications / Janine Kremling.
pages cm. -- (Criminal justice: recent scholarship)
Includes bibliographical references and index.
ISBN 978-1-59332-519-0 (hardcover : alk. paper)
1. Drug abuse--United States--Research. 2. Substance abuse--United States--Research. I. Title.
HV5809.K74 2013
362.29'120973--dc23

2012051269

ISBN 978-1-59332-519-0

Printed on acid-free 250-year-life paper.

Manufactured in the United States of America.

DEDICATION

This book is dedicated to my Professors, family, friends who have supported me all the way and without whom I could not have completed my degree and accomplished my career goal. Your continued support is invaluable.

Table of Contents

List of Tables

List of Figures

CHAPTER 1

Overview of the DUF and ADAM Program: Understanding the Problem

In 1987, the National Institute of Justice (NIJ), in cooperation with the Bureau of Justice Assistance (BJA), implemented a national study tracking drug use prevalence and drug use patterns among arrestees. The program, called the Drug Use Forecasting (DUF) program, started in 12 cities. The DUF program was unique for three reasons: (1) it collected a urine specimen as a validation technique of self-reported drug use; (2) it collected data on drug use prevalence and patterns from arrestees, a population group at high risk for drug use that is not studied consistently across the United States; and (3) it provided local data on drug use prevalence and patterns with the explicit goal of providing policy makers with the necessary information to develop programs that effectively reduce drug use (National Institute of Justice, 1998).

In 1993, the General Accounting Office (GAO) published a report on the strengths and limitations of the three national drug use studies funded by the Federal Government. One of these three studies was the DUF program. The report stated that the major shortcomings of DUF were the use of a non-probability sample, more specifically a judgment-based sample, and a lack of standardization across sites. According to the GAO (1993) these two major shortcomings made it impossible for researchers to generalize findings to the population of arrestees in that specific geographic area. As a result, the DUF data was said to be useless for policy makers, who had as a major goal the development of programs aiming to reduce drug use.

After the evaluation by the GAO, NIJ decided to implement some major changes to the study design. The most important modification was the decision to change the judgment-based sample to a probability sample. In 1998, the name of the study changed from Drug Use Forecasting (DUF) to Arrestee Drug Abuse Monitoring (ADAM) program and the data collection was now standardized at all sites, but the study still used a judgment-based sample. The probability sample was not fully implemented until the latter half of 1999. Similar to DUF, the ADAM study collected data about drug use prevalence and patterns and used urine analysis as a validation technique for self-reported drug use. In addition, ADAM also collected data on drug market activity, drug treatment, and other drug-related issues (National Institute of Justice, 2000).

Due to the implementation of a probability sample, the findings of the ADAM program were now said to be representative of the target population of booked arrestees, allowing researchers to generalize the findings to the general population of arrestees for the geographic area at which the study was implemented. However, the ADAM program was only carried out for approximately three years.

The federal government terminated the ADAM study in January 2003. The high costs of the program and significant budget cuts by Congress have been cited as the major reasons for the termination of ADAM by the National Institute of Justice (Yacoubian, 2004). Originally, Congress had allocated $20 million per year in discretionary money for social science research, but this research money was reduced to $6 million for the year 2004. Thus, the ADAM program became too expensive to continue. Whereas the DUF program costs about 1 million dollars per year, ADAM costs about 8.4 million dollars per year (National Institute of Justice, 2004). The much greater costs of ADAM were due mainly to the greater number of sites where data were collected, the greater amount of time interviewers spent at the facility supervised by a correctional officer, continuous training of the interviewers, modifications to the questionnaire (more extensive and detailed than DUF), and changes in urine specimen processing. All of these changes in the program (from DUF to ADAM) will be described in detail in Chapter Three.

Between 2003 and 2006 no drug use data was systematically collected for the population of arrestees. In 2007, the Office of National Drug Control Policy (ONDCP) revived the ADAM program with its probability sample. However, the ADAM II program was only implemented at 10 sites that were previously part of the ADAM program. These 10 sites represent individual counties in 10 separate states. This is significantly less than the original ADAM program (35 sites) and it is also significantly less than the National Household Survey (NHSDUH) and Monitoring the Future (MTF), both of which are national studies. These differences are important for two reasons: (1) drug use varies by geographic location and (2) arrestees have significantly higher rates of drug use as compared to the general population, which is being studied by the NSDUH and the MTF.

Research has consistently shown that drug use prevalence and patterns vary significantly across geographic locations, and as a result studying only 10 counties in the entire country does not provide sufficient data for drug using behaviors among arrestees (Drug Abuse Warning Network, 2011; Drug Enforcement Agency, 2011; Feucht and Kyle, 1996; Peters, Yacoubian, Baumler, Ross, and Johnson, 2002; Riley, 1997; Yacoubian, 2002). For example, the ADAM data itself shows that the prevalence of certain illicit drugs depends on the geographic location of the site. In 2001, drug test results demonstrated that the percent of arrestees who tested positive for cocaine use ranged from a low of 11.0% in Des Moines, Iowa, to a high of 48.8% in New York. Similarly, the percent of arrestees who tested positive for marijuana ranged from 28.5% in Laredo, Texas, to 54.2% in Minneapolis, Minnesota, and for opiates from 2.0% in Omaha, Nebraska, to 27% in Chicago, Illinois (NIJ, 2001). Emergency department visits for illicit drug-related illnesses were higher in Boston (571 per 100,000), New York (555 per 100,000), Chicago (507 per 100,000), and New York (462 per 100,000) than in the rest o the country (Drug Abuse Warning Network, 2011). Finally, methamphetamine use varies widely with the highest use rates in central and east United States (DEA, 2011).

Additionally, illicit drug use also varies considerably within states. The percent of arrestees who tested positive for cocaine in Texas varied between 20.4% in San Antonio and 45.0% in Laredo. Also, marijuana use in Texas ranged from 28.5% in Laredo to 40.7% in San Antonio (NIJ, 2001). This result demonstrated that Laredo had the lowest number of cocaine users, but the highest number of marijuana users. These within-state differences did not just apply for Texas, but were also apparent for other states including California, Washington, and Florida. Similarly, methamphetamine use varies considerably within states. For instance, methamphetamine use has increased in 21 cities between 2007 and 2009, but decreased in other cities (National Drug Intelligence Center, 2009). These geographic differences in drug use prevalence and patterns demonstrate the importance of collecting data from arrestees nationwide. Collecting data for 10 sites (counties) is not sufficient to provide a comprehensive overview of drug use prevalence and patterns in the United States because these 10 sites are not representative of drug use for other cities and states. As a result, important changes in drug using behaviors may not be discovered at all or they may be discovered only once they have become epidemic.

Second, research has consistently shown that arrestees have substantially higher rates of illicit drug use than the general population (NSDUH) and school children (MTF) (BJS, 2004). In 2009, police arrested 1.6 million people for drug-related offenses (Justice Policy Institute, 2011). Brecht, et al. (2003) estimated that about 65% of arrestees use illicit drugs. By comparison, approximately 8.3% of the general population (as determined by the NSDUH) and about 9.5% of school children use illicit drugs (SAMHSA, 2003; 2008). Thus, the NSDUH and the MTF track illicit drug use nationwide for population groups that use illicit drugs at much lower rates than arrestees. It might be more useful to put a greater focus on population groups who use drugs regularly because regular use of illicit drugs results in great costs for society.

Illicit drug use is very costly for society for several reasons. First, drug use is related to criminal behavior in three ways: (a) users must obtain money for drugs; (b) drugs may have a detrimental effect on the individual's behavior; and (c) as part of the lifestyle and business methods of drug dealers (NIDA, 1990). For instance, studies have found that the rise in crack cocaine was associated with a significant increase in the urban crime rate, especially violent crimes (Bennett, et al., 2008; Grogger and Willis, 2000, Inciardi, 1990). Additionally, violent behavior has been found to be related to the use of other psychoactive substances, such as amphetamines, cocaine, LSD, and PCP (Roth, 1994). Research has also demonstrated that drug users engage in criminal activities, such as burglary and drug sales to obtain monies for drugs (CASA, 2010; Dembo, Williams, Wish, Berry, Getreu, Washburn, and Schmeidler, 1990). Finally, the lifestyle and business practices of drug dealers include violence as part of the interaction between drug dealers, rival gangs dealing drugs, drug runners, and informants (Goldstein, 1987).

Second, drug use contributes significantly to the costs of health care. Specifically, researchers have estimated that almost 50% of all health care costs are related to alcohol and drug use. French and Martin (1996) provide an overview of the cost of illicit drug use for society. According to the authors, these costs can be divided into nine categories. The categories are: "medical services costs; prenatal costs; drug abuse treatment costs; drug-associated disease costs; cost of alcohol, illicit drug and mental health (ADM) comorbidity; crime-related costs; foster care payments; special education and early intervention costs; and costs of Aid to Families with Dependent Children (AFDC) including food stamps" (French and Martin, 1996, p. 454). The Office of National Drug Control Policy has estimated that the total economic costs of illicit drug use are about $193 billion due to a loss of economic productivity, increased health care costs, and costs for the criminal justice system (National Drug Intelligence Center, 2011). The largest share of the losses, an estimated $120 billion, stems from a loss of productivity, including labor participation costs, specialty treatment costs for services provided at the state level and federal level, hospitalization costs, and mortality costs. These costs can only be reduced if drug use decreases. Arrestees are a population group who use drugs at high rates. Therefore it is important to study their drug-using behaviors and implement programs that help decrease their drug use (National Drug Intelligence Center, 2011).

Third, drug use has been shown to be associated with loss of employment, loss of housing, and family disintegration (Henkel, 2011; NIDA, 1990). Each of these consequences constitutes and contributes to the costs of illicit drug use for society. For example, a loss of employment can lead to criminal behavior because individuals who use illicit drugs have to find another way to get money for drugs. They might engage in drug selling or other illegal activities that help them get more drugs. Thus, it is crucial to track drug-using behaviors among arrestees across the nation to be able to implement effective programs. In fact, the National Institute of Drug Abuse (NIDA) suggests that in order to reduce drug use, it is imperative to implement "community-based prevention and treatment programs" (Mueller, 2011; Roth, 1994). If the goal is to decrease the costs of drug use for society and implement local prevention and treatment programs, as proposed by NIDA, then a reasonable approach would include arrestees because they have high rates of illicit drug use (65%), and who then, as a result, contribute greatly to the costs of drug use for society.

MAIN PURPOSE OF DUF AND ADAM

As stated previously, the main purpose of the DUF and ADAM program was and still is (with ADAM II) to guide policy and program implementation at the local level (ONDCP, 2011a). In order to fulfill that purpose it is necessary to know which drugs are being used, how they are being used, and what distinguishes drug users, that is, arrestees who use illicit drugs, at the different geographic areas within the United States. This information can then be used to develop programs and services targeting specific drug users, and programs can be designed and implemented that are tailored towards the needs of a certain community or geographic area. Tailoring the programs to the needs of the community is important in order to be effective in reducing illicit drug use.

CUSTOMIZING PROGRAMS TO THE COMMUNITY

Research on drug abuse treatment has shown that programs specifically customized for a certain individual are more effective in reducing future drug use and related issues, such as recidivism and infectious diseases (Huddleston and Marlowe, 2011; Hammett, Harmon, and Rhodes, 2002; Murphy, Collins, and Rush, 2007). For instance, drug courts have been show to be effective in reducing drug use and criminal activity because they provide individualized treatment based on best practices (Huddleston and Marlowe, 2011). Similarly, programs aiming to prevent the spread of infectious diseases, such as HIV, are also most effective when they meet the needs of the community (Kelly, et al., 1992). Accordingly, it could be expected that programs aiming to reduce illicit drug use among arrestees would be more successful if they would meet the specific requirements of drug users in a certain community. These customized programs might also be more cost effective because they don't waste resources on combating drugs that are not widely used. For example, there is no great need to implement programs targeting methamphetamine users in Florida, because methamphetamine is used rarely in that area. There is, however, a need for such programs in San Diego and other west coast cities where up to 22% of arrestees use methamphetamine. Collecting data locally but expanding the collection sites to be representative of the nation will provide the necessary information to implement customized programs, as a result reducing illicit drug use, recidivism, and the spread of infectious diseases.

The major obstacle, it seems, are the costs associated with such a program. Considering that the total economic costs of illicit drug use are estimated to be approximately $143.4 billion each year (Office of National Drug Control Policy, 2001) and that $50 million is spent by the government via NSDUH to survey a population group that rarely uses drugs (8.2%), it would be reasonable to also implement a nationwide program monitoring illicit drug use among arrestees, 65% of whom use illicit drugs. It is, however, also important to recognize current budget cuts and the economic situation. Thus, this study will examine whether it is possible to implement a study similar to ADAM that is equally effective but less expensive.

PURPOSE OF THE STUDY

As described above, the main reason for the change from a non-probability sample (DUF) to a probability sample (ADAM), and eventually the termination of the study altogether, was the critique of DUF by the General Accounting Office (GAO). A detailed overview of the DUF program and the criticisms of the GAO will be discussed in Chapter Three, but the main conclusion of the GAO was that the sampling procedure of DUF did not allow for the generalization of results to arrestees in general because it is unclear whether the information contained in DUF is valid with regard to drug-using behaviors among arrestees within the geographic areas studied (GAO, 1993).

To date, no one has systematically assessed this question. The purpose of the current study is to examine this issue by comparing the results of the probability sampling used in ADAM to the non-probability sampling of DUF. Specifically, the main research question is whether the non-probability sample of DUF contains drug use information that could be said to be equivalent to the drug use information contained in the probability sample of ADAM.

SIGNIFICANCE OF THE STUDY

The proposed research question is important for two main reasons. First, if the analysis indicates that the DUF and ADAM data do not provide substantially different information, this might help implement a drug abuse monitoring program that observes drug use among arrestees nationwide and provides local data on drug use prevalence and patterns among arrestees. As a result, the research can guide policy development and program implementation aiming to reduce drug use at the local level.

Second, the current analysis is important because the DUF data (data between 1987 and 1999) has only been used by a few researchers. The results of the current study might enable researchers to publish research findings from the DUF data and from both DUF and ADAM over time. This might be especially important for researchers monitoring (a) the relationship between drug use and crime over time; (b) the popularity of certain drugs over time; (c) the introduction of new drugs and how they spread around the country; and (d) the relationship between newly implemented drug laws, policies and drug use prevalence and patterns.

UNIQUENESS OF THE CURRENT STUDY

The current study is also unique. To date, there is little research comparing the data from two different samples, using different sampling strategies, with the purpose of exploring whether these two samples contain substantially different information. Although there is no study assessing differences across the DUF and ADAM data systematically, there is some evidence that the information contained in DUF might not be substantially different from ADAM.

The National Institute of Justice (NIJ) conducted a study comparing drug use outcomes for one site for the DUF and ADAM data (NIJ, 1990). The study used data from the Uniform Crime Report (UCR) as a base rate to assess whether differences in the charge distribution of arrestees lead to biased drug estimates. The results demonstrated that drug use estimates did not appear to be biased. Although the demographic characteristics of the sample were different, the drug use information was similar.

A second study was conducted in Anchorage, AK with the goal of determining whether the male and female arrestees interviewed were representative of the arrestee population at that site (Myrstol and Langworthy, 2005). For this purpose the authors compared the demographic information of the arrestee sample to the demographic information collected via face sheets from all arrestees, including nonrespondents. The authors found that although female arrestees were not sampled in accordance with the probability sampling plan for males, they were more representative of the population of booked female arrestees than the male sample was of the population of booked male arrestees. This is notable because, similar to DUF, female arrestees were selected via a convenience sample. Thus, the results of the study from Anchorage suggest that a non-probability sample might include a similar subset of arrestees as the probability sample. Both studies only examined one site, however.

Additionally, some research has been done examining the equivalence between Internet-based and paper-and-pencil data collection. This research suggests an overall equivalence between these two methods despite differences in the demographic profiles of the two samples (Epstein, Klinkenberg, Wiley, and McKinley, 2001; Krantz, Ballard, and Scher, 1997; Pasveer and Ellard, 1998; Rusticus and Lovato, 2011). This supports the findings of the NIJ (1990) study which also found that demographic differences in the sample did not necessarily result in substantially different drug use information.

Based on these studies, it is possible that the drug use information contained in the DUF sample is not substantially different from the ADAM sample. Thus, the current study attempts to assess the question whether the non-probability sample of DUF provides information about drug use prevalence and patterns that is comparable to that of ADAM in a more systematic fashion by examining all sites that have the same catchment area for DUF and ADAM. This study is possible because both DUF and ADAM examined drug use prevalence among arrestees. Although ADAM had more sites and a more comprehensive interview instrument, there are nine sites that have the same catchment area and contain 14 variables for self-reported drug use and urine test results for the major drugs (i.e., marijuana, cocaine, and opiates), thus providing the necessary information for this assessment.

This study is also unique because it uses equivalence testing to examine the proposed research question. Equivalence means that there are not substantial differences (Rogers, et al. 1993). Equivalence testing is an analysis strategy widely used among clinical researchers to assess whether two different drugs/treatments produce a comparable outcome. Equivalence testing assumes that two different drugs/treatments will always result in some differences, but these differences might not be of practical and/or clinical importance.

Similarly, it can be assumed that two different samples will result in different outcomes. The crucial question, and the significant research question of the current study, is whether or not the sampling design and procedures used by ADAM resulting in an approximation of a true probability sample produced results that are substantially different from DUF. Equivalence analysis is applied to assess this research question.

BRIEF OVERVIEW OF EQUIVALENCE TESTING

The method employed in this study is the confidence interval method first described by Westlake (1981). Rogers, et al. (1993) introduced the method to the field of psychology. The main idea of this method is to calculate confidence intervals for the proportions for the DUF and ADAM drug estimates and conduct a traditional hypothesis test and an equivalence test simultaneously. The outcome of these two tests will show whether the DUF and ADAM data are substantially different or whether they are sufficiently similar to be considered equivalent. Substantially different means that drug use estimates contained in DUF and ADAM are statistically different in the traditional null hypothesis test and not equivalent in the equivalence test. Overall, substantially different is defined as a difference of 20% or more between the drug use estimates of DUF as compared to ADAM. Equivalence is present if the drug use estimates are statistically significant in the equivalence test and not significant in the traditional null hypothesis test. Equivalence does not mean "exactly the same" rather it means the "absence of a meaningful difference" (European Medicines Agencies, 2000; Rogers, Howard, Vessey, 1993; Allen and Seaman, 2006; Tryon and Lewis, 2009). Thus, overall, equivalence is said to exist if the difference between the drug use estimates in DUF and ADAM is less than 20%. The exact method is described and demonstrated on an example in Chapter Four.

POSSIBLE OUTCOMES OF THE ANALYSIS

Four possibilities exist with regard to the outcome of the current analysis:

1) The drug use estimates of the DUF and ADAM samples are equivalent (Eq).
2) The drug use estimates of the DUF and ADAM samples are different (D).

3) The drug use estimates of the DUF and ADAM samples are different and equivalent (D&Eq).
4) The drug use estimates of the DUF and ADAM samples are not different and not equivalent. They are statistically indeterminate (ND&NEq).

Possibilities one and two are fairly straightforward. First, the drug use estimates of the DUF and ADAM samples can be said to be equivalent if the drug use proportions are statistically equivalent and not statistically different. Second, the drug use estimates of the DUF and ADAM samples can be said to be different if the drug use proportions are statistically different and not statistically equivalent. Third, the drug use estimates of the DUF and ADAM samples can be said to be equivalent and different if the drug use proportions are statistically different but statistically equivalent. In this case the researchers suggest that the data might not be substantially different. Rather the differences can be said to be trivial (Allen and Seaman, 2006; Rogers, et al. 1993). Fourth, if the drug use estimates of the DUF and ADAM samples are not statistically different and not statistically equivalent no conclusions can be drawn with regard to the question whether there exist substantial differences. The results would be ruled indeterminate (Rogers, et al. 1993; Tryon and Lewis, 2009). For an easier overview Table 1.1 demonstrates the four possibilities.

Table 1.1. Possible Outcomes of the Analysis

		Equivalent	
		No	Yes
Different	Yes	2 (D)	3 (D & Eq)
	No	4 (ND & NEq)	1 (E)

Before the question about the equivalence of the DUF and ADAM data can be assessed, it is critical to examine why it is important to study illicit drug use in general and why ADAM and DUF were crucial for researchers, communities, and policy makers in their assessment and efforts to reduce drug use. Thus, Chapter Two discusses the importance of studying illicit drug use in general, followed by a description of the importance of the DUF and ADAM programs for communities, policy makers and researchers, and why the original termination and subsequent revival of the study with only ten sites in 2006 constitutes a major loss and an obstacle to the goal of reducing drug use. Chapter Three lays out the methodology of DUF and ADAM and discusses the major criticisms of the GAO (1990) on the DUF program and the specific changes that were made by the National Institute of Justice to improve the program and develop a dataset that could be generalized to the greater population of booked arrestees within the geographic area studied. Chapter Four explains the analytical plan for the research question and the statistical analysis employed and a specification of the variables included in the current study. Chapter Four also presents the descriptive statistics for the demographic characteristics of the DUF and ADAM samples and drug use information. Chapter Five presents results for the equivalence analysis determining the comparability of the DUF and ADAM data with regard to drug use prevalence and patterns. Finally, Chapter Six concludes by summarizing and discussing the results of the analysis and providing implications of the results for future research.

CHAPTER 2

Why Is It Important To Study Illicit Drug Use?

BUDGET SPENT ON COMBATING DRUG USE

Illicit drug use is a problem of high priority in the United States (ONDCP, 2011b; Reuter, 2006). One indicator of the importance of reducing illicit drug use is the amount of monies spent by the government on decreasing illicit drug use as compared to other expenditures. The Office of National Drug Policy (ONDCP) estimates the costs of drug expenditures of the federal government. Until 2002, the ONDCP combined drug-targeted (e.g., domestic and international enforcement) and drug-related (e.g., prevention and treatment) expenditures in their drug expenditure estimation. The drug-related expenditures also included substance abuse and rehabilitation research (ONDCP, 2002). Using this comprehensive approach, the ONDCP estimated that the government would spend $19.2 billion on the national drug control budget for the fiscal year 2002 (ONDCP, 2002). This estimate decreased to $12.9 billion when expenditures associated with the consequences of drug use (e.g., cost of incarceration) were excluded (ONDCP, 2002).

For the year 2010, the federal government has provided a budget of $15.1 billion (ONDCP, 2009). This budget includes funding for treatment, prevention, domestic law enforcement, interdiction, and international counterdrug support. In comparison, the U.S. Department of Justice is allocated a total budget of $26.5 billion, and the budget for the U.S. Department of Education is $46.7 billion. Also, the budget allocated to combating illicit drug use is greater than the total budget for the U.S. Department of Commerce, which receives only $12 billion.

In sum, the federal government allocates a significant amount of money to reduce illicit drug use.

In addition to the fact that the government spends a considerable amount of money on combating illicit drug use as compared to other expenditures, these expenditures have significantly increased within the past four decades. Specifically, the budget of the Drug Enforcement Agency (DEA) increased from $65.2 million in 1972 to $15.1 billion for 2010 (ONDCP, 2011). At the same time, the number of total employees increased from 2,775 in 1972 to 10,891 in 2006 (Drug Enforcement Agency, 2007a). State and local agencies also spend a sizable amount of their budgets on drug law enforcement. For example, New York and California each spend about $1 billion per year on law enforcement efforts related to the prohibition of marijuana use alone (Drug Reform Coordination Network, 2005). Overall, drug law enforcement receives a sizable amount of funding at the federal, state, and local levels.

POLICIES AND PROGRAM COMBATING DRUG USE

The importance of combating illicit drug use and abuse is also illustrated by the fact that the government has implemented a considerable number of policies and programs targeting illicit drug use. The policies and programs are aimed either at drug supply reduction or drug demand reduction. An example of a program targeting drug supply would be the Organized Crime Drug Enforcement Task Force, which combines the expertise and resources of all federal agencies involved in drug law enforcement (including the FBI, the Bureau of Immigration and Customs Enforcement, the Bureau of Alcohol, Tobacco, Firearms and Explosives, and the U.S. Marshals Service) with the goal of combating major drug trafficking and money laundering (DEA, 2007b). An example of a drug demand reduction program would be the Drug-Free Schools and Communities Act administered by the U.S. Department of Education. The proposed goal of the act is to educate school children of the dangers of alcohol and illicit drug use and to prevent such illicit drug use (Office of Safe and Drug Free Schools, 2006). Additionally, the National Youth Anti-Drug Media campaign aims to prevent drug use via the national campaign "Above the Influence" and community-based campaigns. Similarly, the Anti-Meth Campaign focuses mainly on rural and sub-rural areas and targets young adults between the ages of 18 and 34 (ONDCP, 2010).

IMPORTANCE OF STUDYING DRUG USE FOR RESEARCHERS

Illicit drug use is of great importance not only for law enforcement but also for researchers. There is a large amount of research examining drug use prevalence and patterns across different population groups, predictors of drug use, evaluation of drug prevention programs (school- and community programs) and drug treatment programs, the cost-effectiveness of treatment programs, and the effectiveness of drug courts. This section reviews the research relevant to the current study – drug use prevalence and patterns as shown in the major national drug studies. This is important because it will demonstrate how important DUF and ADAM were for drug researchers and why it is imperative for the advancement of drug researchers and the implementation of effective drug reduction programs to systematically monitor drug-abusing behaviors among arrestees across the United States.

ONGONG NATIONAL STUDIES OF DRUG USE PREVALENCE AND PATTERNS: WHY ADAM SHOULD BE IMPLEMENTED NATIONWIDE

The government has collected drug use data via self-report surveys administered to nationally representative samples of households (National Survey on Drug Use and Health) and youths (Monitoring the Future) for more than thirty years. Although both studies survey a nationally representative sample of individuals, they rely on data sources with low credibility (self-report) and they survey population groups who are not heavy substance users. Additionally, the Substance Abuse and Mental Health Administration (SAMHSA) sponsors the Drug Abuse Warning Network (DAWN) program, which monitors drug related emergency room visits. Next, a brief overview of the NSDUH, MTF, and DAWN will be provided for a better understanding of their drug estimates and why these surveys have not played much of a role for drug researchers and policy makers.

THE NATIONAL SURVEY ON DRUG USE AND HEALTH (NSDUH)

The National Survey on Drug Use and Health (NSDUH), formerly known as the National Household Survey on Drug Abuse (NHSDA), was implemented by the Federal Government as an annual survey in

1971. The NSDUH targets a representative sample of the civilian, non-institutionalized population aged 12 years and older via face-to-face interviews in all 50 states and the District of Columbia (Department of Health and Human Services, 2011). Thus, the survey intentionally excludes persons who are institutionalized, such as persons in jail, prison, or mental hospitals. The survey also excludes persons who have no fixed address (i.e., homeless and transient persons), and military personnel.

The NSDUH inquires about alcohol and drug use for the following drug classes: alcohol, marijuana, cocaine, heroin, hallucinogens, inhalants, psychotherapeutics, and tobacco (SAMHSA, 2005). The questionnaire first asks participants whether they have ever used any of these drugs. If the participants report drug use for any of these drug classes, the interviewer continues with more detailed questions for each drug used, including the last time used and age of first use. The survey also includes a number of questions regarding demographic characteristics, such as age, gender, pregnant women, education, and employment. Additionally, participants provide information about previous drug treatment, need for drug treatment, needle sharing, experienced consequences of drug use, and their criminal history (SAMHSA, 2005).

Over time, the NSDUH underwent a number of methodological changes, which included changes to the sampling method, the interview method, and the questionnaire (Gfroerer, Eyerman, and Chromy, 2002). The latest change in the sampling procedure was implemented in 2005. The NSDUH now employs a multi-stage probability sampling design consisting of three phases: (1) Stratification of States into 900 State sampling regions; (2) Selection of 48 census tracts per State sampling region, (3) Selection of area segments (census blocks). From these area segments, four samples are drawn—one for each quarter of the calendar year—allowing for continuous data collection. For each area segment, a listing of the addresses are obtained and sampling units selected. Finally, the interviewer randomly selects the sample person via a computerized procedure (SAMHSA, 2007).

Changes to the NSDUH also include modifications to the interview method. Until 1998, the survey was conducted via paper and pencil method. In 1999, the paper and pencil method was changed to a computer-assisted method to increase the response rate. Computer-assisted methods have been shown to result in higher reports of drug use, typically attributed to the higher degree of confidentiality.

Respondents stated that they were more truthful because the interviewer would not know about their drug use (SAMHSA, 2000).

Several other methodological changes were also implemented (SAMHSA, 2007). The name of the program was changed from National Household Survey to National Survey on Drug Use and Health (NSDUH). More importantly, survey participants now receive a payment of $30, which has substantially increased the response rate. Additionally, the data quality control procedure (including training sessions for staff, higher degree of supervision of the interviews, evaluation of interviewers) was improved during 2001 and 2002. The sampling weighting procedures of the 2002 survey are based on the 2000 decennial data, whereas the sampling weighting procedures for previous years are based on the 1990 decennial data (SAMHSA, 2007).

In 2007, SAMHSA conducted a study examining the impact of the methodological changes in the NSDUH. Results showed that the response rate had improved significantly, which was probably due to the $30 incentive and the implementation of the computer-assisted interviewing method. Also, the reported drug prevalence rates were significantly higher, which was attributed to the increased quality control procedures and the computer-assisted interviewing method (SAMHSA). The results of the analysis suggest that the 2002 data should not be compared to the results of previous years because the methodological changes had a significant impact on the study outcomes. This also implies that researchers interested in changes in drug prevalence would not be able to use all data between 1971 and 2002 (SAMHSA, 2007).

Although the NSDUH is representative of the general population of the United States, it has some major limitations. First, the NSDUH excludes population groups that have been shown to be at high risk of drug use, specifically, institutionalized persons and homeless persons. As described previously, there is considerable evidence that persons who are institutionalized in jails, prisons, and mental hospitals have much higher rates of illicit drug use than the civilian, non-institutionalized population. These civilian, non-institutionalized persons are likely only occasional users who mostly use marijuana, and increasingly prescription medication, but who do not provide information about recent changes in drug paraphernalia and the introduction of new or modified drugs (i.e., methamphetamine, crack cocaine). It is, however, the regular use of hard drugs (such as cocaine, heroin, methamphetamine) and the introduction of new and/or modified drugs and changes in drug prevalence and patterns that are the most crucial for researchers, policy makers, local law enforcement, and communities in their attempts to develop programs that effectively reduce drug use. The occasional user of marijuana is not the major problem.

Second, the NSDUH relies on self-report data. As will be described in more detail in the next section, self-report data has been shown to greatly underestimate drug use. This limitation is a major one, as it may lead to conclusions about illicit drug use that are not correct. The use of incorrect information about drug use by law enforcement and policy makers can result in the implementation of policies that are determined to be ineffective when evaluated. The implementation of ineffective drug combating strategies is a waste of resources that could be used more effectively elsewhere.

Third, the NSDUH assesses geographic differences in drug use for the four major areas: the West, Midwest, Northeast, and South. The results suggest that overall drug use is quite similar across the United States. In 2010, the West had an overall drug use rate of 11.0%, the Midwest had an overall drug use rate of 8.2%, the Northeast had an overall drug use rate of 9.4%, and the South had an overall drug use rate of 7.8%. The NSDUH does not track changes and differences for individual drugs for these geographic regions. This is important because monitoring local areas can be of great help for local law enforcement and communities because they are able to target hot spots of drug sellers and users. As mentioned earlier, it is also important for the implementation of local programs providing drug abuse treatment and for the reduction of the spread of infectious diseases. This type of information cannot be provided by the NSDUH because the civilian, non-institutionalized population very likely does not "hang out" at such hot spots and can therefore not provide information about it. Yet, it is these "hot spots" and the persons who are dealing and using drugs that cause the major problems for communities.

MONITORING THE FUTURE (MTF)

Another national survey on drug use and abuse is Monitoring the Future (MTF). The MTF is a longitudinal survey implemented in1975 with the intent to collect data about attitudes towards drugs and drug using behaviors among high school children in the United States

(Bachman, Johnston, O'Malley, and Schulenberg, 2011). The goal is to collect data from a nationally representative sample of high school students within the 48 contiguous states.

The MTF surveys approximately 50,000 high school children in the spring of each year. The survey includes approximately 110-120 public high schools and 15-20 private high schools. The data is collected via a stratified, multi-stage sampling procedure: (1) selection of the geographic area, (2) selection of schools in the selected geographic area, and (3) selection of participants in each school. The survey is then administered during regular class periods.

The survey asks extensively about the use of licit and illicit drugs including marijuana, sedatives, tranquilizers, hallucinogens, amphetamines, cocaine, heroin, inhalants, steroids, alcohol, tobacco, stimulants, diet aids, and creatine. The survey also asks the participants about attitudes towards drugs, consequences of their drug use, and whether they believe that they could stop or reduce their drug abusing behaviors (Bachman, et al., 2011). Additionally, the MTF asks a series of questions about demographic characteristics, including race, gender, parental education, and questions about school performance and satisfaction with school (Bachman, et al., 2011).

The survey's greatest value lies in its ability to track changes in drug use over time, because the survey is administered repeatedly to the same segments of the population in private and public high schools (8th, 10th, 12th graders, college students, and young adults). As a result, the MTF allows researchers to assess changes in drug use prevalence and patterns in four areas: (1) Period effects (changes across all age groups for a certain year), (2) Age effects (changes in drug use for all panels), (3) Cohort effect (differences among cohorts throughout the life cycle), and (4) Differences attributable to differences in the environment (e.g., high school, employment) and changes in life (e.g., marriage, military, parenthood) (Johnston, O'Malley, Bachman, and Schulenberg, 2007).

The MTF, however, also suffers from a number of limitations. First, similar to the NSDUH, the survey only collects data via self-report surveys. Second, school drop-outs are not captured in the data. This is problematic because research has demonstrated that drug users are overrepresented among high school drop-outs (ADAM, 2000). Third, some schools may decline participation. These schools might differ from schools that are participating in ways that could bias the sample. Fourth, the MTF does not provide representative data for local areas (only at the national level). Combating drug abuse, however, is very important for local law enforcement and communities. The data from the MTF is not especially helpful for such purposes.

As stated earlier, both the NSDUH and the MTF are quite expensive surveys that accomplish very little with regard to providing useful information about drug use, especially hard drugs such as powder cocaine, crack, heroin, and methamphetamine. They are also not helpful in determining drug use hot spots and the emergence of new drugs. It might be time to examine whether part of the funding that goes to these two surveys can be redistributed to a national survey that studies drug use among arrestees. Approximately 70% of jail inmates (BJS, 2002), 83% of state prison inmates (BJS, 2004), 27% of parolees, and 29.9% of probationers (SAMHSA, 2011) reported drug dependency issues, but only about 10.1% of school children, 21.5% of young adults between 18 and 25, and 6.6% of persons 26 years of age or older have used illicit drugs (SAMHSA, 2011). Considering the fact that jail inmates and prisoners are using illicit drugs at much higher rates than the general population, it would be more useful to examine arrestees than school children and non-institutionalized adults with regard to illicit drug use, drug market activity, and criminal behavior associated with drug use.

DRUG ABUSE WARNING NETWORK (DAWN)

A third national data collection tool that provides information about the extent of illicit drug use is the Drug Abuse Warning Network (DAWN). DAWN, also sponsored by SAMHSA, was first implemented in 1988 as a program that monitors drug-related emergency room visits and drug-related deaths (SAMHSA, 2008). In 2003, a new methodology was implemented with the goal to improve the quality and utility of the monitoring system. Changes were made to the following areas: sample, target population, geographic boundaries, definition and method of finding dawn cases, data content, and supervision of data quality. Due to the changes in the methodology, SAMHSA data from 1988 until 2002 cannot be compared to the data collected in 2003 and later years (SAMHSA, 2008).

The new DAWN collects data from a sample of hospitals representative of 50 states and the District of Columbia. Eligible hospitals are short term hospitals, general hospitals, and non-federal hospitals operating 24-hour emergency departments (ED's). The new

DAWN includes all types of drug-related ED visits regardless of the patient's intent or age. Medical charts of ED visits are retrospectively reviewed, and cases that meet the criteria are selected. Demographic information and self-reported drug use information obtained from the patient is recorded. DAWN collects information about all types of drugs, including illegal drugs, alcohol, alcohol in combination with other drugs, dietary supplements, prescription and over-the-counter drugs, and non-pharmaceutical inhalants. The new DAWN program also includes toxicology results as a validation technique for the self-reported drug use. Before 2003, toxicology results were not used as a confirmation method (SAMHSA, 2008).

The greatest strength of the new DAWN program is that it uses toxicology reports to confirm self-reported drug use and that it is a representative sample of the complete United States. Despite the changes in the methodology, there are two major limitations to the DAWN program. First, the data provided by DAWN is an incidence measure of the consequences of drug use (which could be first time or long-term drug use) and does not provide information about the prevalence of drug use. Second, the data provides information about emergency room incidents, not the number of patients, because one patient could be treated in the emergency room several times and it would be recorded as a separate case each time. Thus, it is not clear how many people use drugs, how often they use drugs, where they buy their drugs, and other information crucial for researchers and policy makers in combating illicit drug use.

ADVANTAGES OF DUF AND ADAM OVER OTHER NATIONAL SURVEYS

The DUF and ADAM program not only complemented the NSDUH, MTF, and DAWN, but also had a number of advantages over the other three surveys that greatly benefited researchers, policy makers, local law enforcement agencies, and communities in ways that none of the other three surveys could. The exact methodology of DUF and ADAM will be explained in Chapter 3. This section will focus on examining the advantages of DUF and ADAM as compared to the NSDUH, MTF, and DAWN and why the DUF and ADAM program were so important for researchers, law enforcement, and policy makers.

The three major advantages of the DUF and ADAM program were: (1) They assessed drug use among arrestees - a population group shown to be at great risk of drug use, (2) they used bioassays to validate the self-report data, and (3) they provided local data for every quarter of the year for a period of 15 years.

As discussed in the previous chapters, DUF and ADAM focus on a very different population group than the other three national surveys. Specifically, they focus on arrestees, a subpopulation with a significantly higher rate of drug use than the general population or school children. For instance, the Bureau of Justice Statistics has found that 56% of state and 50% of federal prisoners have substance abuse issues (Mumola and Karberg, 2006). These rates are significantly higher than in the general population. The results from the National Survey on Drugs and Health (NSDUH) show that only about 10% of people 12 years and older have substance abuse issues (SAMHSA, 2011). This means that prisoners are about 5 times more likely to use illegal substances than the general population.

Another important difference is that psychotherapeutic drugs are among the preferred drugs used in the general population. Psychotherapeutic drugs include anti-anxiety drugs (i.e., Xanax, Valium), antidepressant drugs (i.e., Zoloft, Proxil), antimatic drugs (i.e., Eskalith), and antipsychotic drugs (i.e., Thorazine). These psychotherapeutic drugs are not the preferred drugs of prisoners and arrestees. In sum, drug use behaviors among arrestees, prisoners, and the general population vary significantly. Studying drug use among the general population and not among arrestees and other population groups at high-risk of illegal substance use will result in misleading conclusions about drug use in the United States. Overall, researchers estimate that the national surveys miss the vast majority of the drugs that are being used. For instance, Kleiman (2004) states that the NSDUH accounts for only about 30 metric tons out of about 300 metric tons of the cocaine consumption. Thus, when policy makers implement strategies aiming to reduce cocaine demand based on the household survey, they are relying on information from the wrong population group.

Until 2003, the DUF and ADAM programs were the only national drug study that allowed researchers to assess the validity of self-reported drug use. This is important because the main criticism of self-reported behaviors pertains to the question of how valid the results are. Validity for this purpose refers to the question "whether the data recorded by the researcher accurately reflect the phenomenon under investigation" (Harrell, 1985). Stated differently, do people answer questions in self-report surveys truthfully?

Since 2003, the DAWN program also uses bioassays to confirm self-reported drug use. The NSDUH and the MTF still rely on self-report only. This is a major limitation. Thus, not only do these two expensive national surveys assess drug use among the general population, which only accounts for a minimal part of the drug use activity, but these surveys also do not validate the self-report data via objective validation techniques.

VALIDITY OF SELF-REPORT DATA

Self-report surveys have a long history and the validity of the obtained data has been suspect to many researchers because the collected data might be erroneous (Stone, Turkkan, Bachrach, Jobe, Kurtsam, and Cain, 2000). As early as the late 1940s, researchers found systematic biases in self-reported behaviors, probably due to incorrect information provided by the study participants (Stone, et al., 2000).

Although early research on the validity of illicit drug use suggested that respondents reported drug use fairly accurately, more recent improvements in technology have allowed researchers to employ more sophisticated validation techniques which have led to a different conclusion (Harrison, 1995). The results of these more recent studies suggest that self-reported drug use is not reported as accurately as thought. Rather, respondents are likely to underreport drug use. For example, Fendrich, et al. (1999) found that respondents from a high-risk community sample heavily underreported cocaine and heroin use. They found that only 20% of cocaine-positive respondents also admitted to current use of cocaine. Similarly, Appel, et al. (2001) found that only 26% of the respondents from a sample of homeless and transient persons in New York who tested positive for cocaine use also reported their drug use.

There is, however, some research suggesting that the underreporting of self-reported drug use might not be as large as believed. A recent SAMHSA study suggests that there is a high degree of agreement between self-reported drug use and urine test results among the general population. Specifically, for marijuana use there was 89.9% agreement and for cocaine use there was 98.5% agreement

(Harrison, Martin, Enev, and Harrington, 2007). Similarly, Sutton et al. (2011) found that self-reported criminal activity using the life event calendar method was highly predictive of actual criminal activity among prisoners. One criticism of the study is, of course that only criminal activity that was actually caught by the police could be used to validate responses. There is, however, a great dark figure of crime, and as a result there is no way of knowing the "real" criminal activity. Stated differently, prisoners might have reported crimes for which they were caught, but not the "real" amount of crimes they have committed. Other studies show that the agreement between self-reported delinquency (as measured by self-reported arrests) and official data (arrest records) is between 50% and 83% (Hindelang, et al., 1981; Hindelang & Krohn, 2000). Hindelang and Krohn (2000) conclude that these agreement rates are reasonably high. It is, however, impossible for researchers to know whether 50% or 80% of their study participants told the truth unless the data is verified via objective measures, such as official arrest records, criminal records, or urine analysis. It appears that it would be of great importance to know if only 50% of the subjects gave true answers. Thus, although some research suggests that self-report data can be a valid measure of delinquency and drug use, there is a substantial amount of research suggesting otherwise.

A number of researchers have also demonstrated that the accuracy of self-reported drug use depends on the type of drug and the population subgroup. The more stigmatized the drug, the less likely respondents report its use (Harrison, 1992, 1995; Mieczkowski, et al., 1991). Miller and Kuhn (2011) demonstrated that self-reports of marijuana use is very low. Similarly, use of cocaine among inmates of a correctional institution was only 42% (Vignali, Stramesi, Vecchio, and Groppi, 2011). Using data from the DUF program, Harrison (1992) found that arrestees most accurately report use of opiates (60%), followed by marijuana (55% concordance). Arrestees were least likely to report the use of cocaine (50% concordance) and amphetamines (40%). This also holds true for other population subgroups. A study by Hser, Maglione, and Boyle (1999) suggests that self-reported drug use among emergency room patients (ER) and patients with sexually transmitted diseases (STD) was more accurate for marijuana as compared to any other drug. Hser', et al., (1999) study also demonstrates that different population subgroups are more or less likely to accurately report drug use. ER patients and STD patients, as compared to arrestees, were less likely to report drug use overall, but especially the use of hard drugs. The reason may be that ER patients and STD patients perceived the social stigma associated with illicit drug use to be greater than arrestees, who were already stigmatized due to the arrest itself and admitting using drugs may be a minor issue compared to their arrest (Hser et al., 1999). More recent research by Golub, et al. (2005) has found that disclosure rates also vary across geographic locations.

Epidemiological research suggests that people even fail to disclose illicit drug use in situations where a failure to do so might potentially harm them. For instance, Marroun, et al. (2011) found that women don't accurately report prenatal cannabis use. Similarly, Tassiopoulos, et al. (2004) found that 34.2% of out-of-treatment heroin users did not admit that they were also using cocaine. In an emergency situation, failure to disclose the use of an illicit drug may have a negative impact on medical care. It appears that the stigma associated with drug use is perceived to be so strong that it is kept secret even in life-threatening situations. If individuals do not disclose drug use in situations where their health is at stake, why would they disclose such behaviors to an interviewer?

Though there is considerable evidence that drug use is generally underreported, there are exceptions. Vignali, et al. (2011), demonstrated that study participants (e.g. inmates in a correctional facility) would over-report heroin use in hopes of being admitted to at drug treatment program. Also, persons who have either recently been admitted into a rehabilitation program or who have recently finished a drug treatment program report drug use fairly accurately. A study by Hindin, et al. (1994) using data from persons entering a residential drug treatment program found that 89% percent of the respondents who tested positive for cocaine use also reported such use. The concordance rate for heroin was even higher, with 96% of the individuals who tested positive for heroin use also reporting such use.

In some instances, over-reporting can occur. For instance, arrestees may over-report drug use if they believe that it may increase their chances to enter a drug treatment program instead of going to jail (Hser, 1999). Arrestees may also over-report drug use if they have been arrested for a violent crime and anticipate a long prison sentence. Drug use in this case may serve as a mitigating circumstance and make them appear less culpable (Nurco, 1985). Other population subgroups may also over-report drug use. For example, adolescents and college students might over-report drug use to make their behavior patterns fit the behavior that is accepted among their peers. Johnston and O'Malley (1997) found that adolescent males as well as male and female college students reported fewer incidents of drug use in a follow up interview. The researchers concluded: "the 'revised' may well be the more accurate number, and the answers given at earlier ages . . . may be inflated" (p. 78). Both under-reporting and over-reporting can lead to inaccurate conclusions about drug use. DUF and ADAM were unique because they validated self-reported drug use with drug tests, therefore providing a more objective measure of drug use among arrestees. Without the validation of self-reported drug use via bioassays, researchers and policy makers are left guessing about how accurate the collected data is.

EXPLANATIONS FOR THE FAILURE TO REPORT BEHAVIORS TRUTHFULLY

Several explanations have been advanced to explain the under-reporting of drug use. Threats to the validity of self-report data are typically said to stem from the failure to remember events accurately and the failure to report behaviors truthfully (Harrell, 1985). For researchers studying the prevalence and patterns of illicit drug use and other stigmatized behaviors, threats to validity stem mostly from the second category—the failure to report behaviors truthfully. To further explore this question, researchers have proposed a number of theoretical explanations of why respondents may not answer questions in self-report interviews truthfully even though they are assured that the answers are kept fully confidential or that the answers are anonymous (SAMHSA, 2009; Sloan III, Bodapati, and Tucker, 2004). One of these explanations is the "Social Desirability Theory."

The purpose of the self-report surveys is typically very obvious to the respondent, and as a result it is rather easy for the respondents to manipulate the answers (Cook and Selltitz, 1964). The underlying assumption of the social desirability theory about human nature is that humans are social beings and that their behavior is oriented on the behaviors of others (Weber, 1968). Accordingly, social desirability theory advances the thesis that persons will respond to questions in a way that is consistent with social norms, expectations, and "socially desirable traits" (Zerbe and Paulhus, 1987, p. 250). In other words, people will respond to questions in a way that places them in a good light. Researchers have found support for the social desirability theory in a variety of topics. The social desirability thesis first received empirical support from a study conducted in 1953 by Edwards. Edwards (1953) found that there is a relationship between what survey respondents believed to be the socially desirable answer and the answer given by these respondents. Additionally, Willis and Schechter (1997) examined how study participants felt about what the interviewer might think about them and react towards them if they disclosed their drug abusing behaviors. Their study demonstrated that individuals were very concerned about interviewer reactions and preferred not to talk about their drug using behaviors because they felt that they were being judged.

There is considerable evidence that self-reported behavior is biased towards normative behavior (Lord, et al. 2005). For instance, the Magazine Audience Group by Crossley Incorporated (1941) examined the validity of self-reported educational attainment. The results of this study showed that individuals exaggerate their educational attainment regarding graduation from the various educational institutions (e.g., grade school, high school, and college) (Parry and Crossley, 1950). In a study about redeeming war bonds, Hyman (1944) found that individuals were likely to deny that they had redeemed war bonds. The denial of redeeming war bonds was greater for individuals with a higher income status. Presser (1990) found that people over-reported voting behavior and the attendance of religious services. Fear of embarrassment is an important factor in people's decision to tell the truth (SAMHSA, 2009).

To control for a social desirability bias, researchers have included measures in their questionnaires meant to provide information about whether the participant may have altered his/her answers because he wanted to give the "correct" answer. The most commonly used measure for this purpose is the Marlowe-Crowne Social Desirability Scale (Crowne and Marlow, 1964). Unfortunately, the purpose of this scale is quite transparent to the respondents and respondents may answer the questions of the scale honestly but not the questions about drug use. As a result, researchers might interpret the self-reported behavior as accurate when it is really not (Richter and Johnson, 2001).

Threats to the validity of self-reported drug use also arise from the fact that people may not recall events accurately. Even when the respondent is motivated to provide accurate information about self-reported drug use or other behavior, they may simply not remember which drug they used at what point in time or how often they used it. The drug use itself may distort their recall of events (Catania, 1993). Harrell (1985) suggests that general behaviors are easier to recall than specific behaviors. For example, drug users might recall accurately having used a certain drug in the last few days, but may not remember how often they have used the drug or how much of the drug they used.

Additionally, more recent events are recalled more accurately than events that occurred a while ago (Harrell, 1985). This is also referred to as recall decay (Johnson, et al. 1997). A study by Roberts, et al. (2005) suggests that individuals underestimate violent behaviors within a 1-3 year time period using a life-events calendar. This type of calendar was also used by the ADAM study to assess drug use prevalence and patterns within the past 12 months. Recall decay is not only a problem for researchers studying drug using or criminal behaviors but also for researchers in other fields including medicine. Studies have found that individuals will significantly underestimate injury rates if asked more than 2 months after the incident (Jenkins, Earle-Richardson, Slingerman, and May, 2002). Considering that many people cannot accurately remember which injuries they suffered or violent behaviors, recall decay is a serious threat to the validity of drug use data.

Additionally, individuals may believe that a certain event occurred more recently than it actually did. This is called forward telescoping. Individuals might fairly accurately recall whether they ever used a certain drug, but they might not be accurate in recalling in which year or during what time period the drug was used (Johnson and Schultz, 2005). Telescoping is especially a problem when an event occurs regularly and is not salient in the sense that it would be remembered easily (Magnusson and Bergman, 1990). Many drug users use drugs regularly, some several times a day. It is unlikely that these regular drug using behaviors are particularly salient, and as a result it cannot be expected that drug use is being recalled accurately.

IMPLICATIONS OF THE LACK OF REPORTING FOR RESEARCHERS AND PLICY MAKERS

The implications of the lack of complete reporting are manifold. For instance, it has implications for researchers who are drawing conclusions about the relationship between drug use and crime, changes in drug use prevalence and patterns, and other issues. A lack of truthful reporting will necessarily lead to inaccurate conclusions. If the goal of research is to advance the knowledge in a certain field publishing results based on incorrect information would be counterproductive. The most important implication of a lack of truthful reporting behaviors of drug use probably pertains to the implementation of policies meant to reduce drug use, provide treatment to drug users, and prevent the spread of infectious diseases, including HIV and Hepatitis (Des Jarlais, 1998). If programs are based on invalid data, then the effectiveness of these programs might be quite low. Additionally, a lack of truthful reporting (mostly under-reporting) might lead to the conclusion that drug abuse treatment programs are not needed in a certain community. As a result, drug users will not receive treatment and continue to use drugs, increasing the likelihood of the spread of infectious diseases and criminal behavior associated with drug use.

Due to the great likelihood that self-reported data is inaccurate, especially for sensitive and stigmatized behaviors (such as illicit drug use), researchers need to validate data whenever possible. The ever-present issue is what measure could be used to assess the truthfulness of the provided information.

VALIDATION METHODS FOR SELF-REPORTED DRUG USE

In the case of illicit drug use, researchers have a variety of validation techniques available, including the analysis of urine specimens, hair samples, sweat, or saliva (Vignali, et al., 2011; Cone, 1997). Each of these methods is unique and provides different types of information. Each method also has its strengths and weaknesses. The usefulness of a certain drug testing method depends on a number of factors, which all relate to the accuracy with which a certain method is able to detect drugs in biological fluid or tissue (Cone, 1997). The factors influencing the usefulness of a testing method are: (1) sensitivity; (2) specificity; and, (3) accuracy. Sensitivity refers to the "least amount of detectable drug," meaning that the more sensitive the test, the lower the concentration of the drug that can be measured (Cone, 1997, p. 109). Specificity refers to "how selective the assay is for the drug" or the ability of the test to distinguish between different drugs (Cone, 1997, p. 109). The higher the specificity of a test, the more accurately it can determine the presence of a certain drug. The most common method utilized in drug use research is urine testing. Both DUF and ADAM employed this method. Research has suggested that, at the current time, it is the most accurate drug screening method for recent drug use (2-4 days) as compared to hair, saliva, or sweat (Mieczkowski and Newel, 1997). Other research, however, has demonstrated that hair testing is superior to urine analysis for detecting occasional drug use because hair can still include traces of drugs after several months (Polla, Stramesi, Pichini, Palmi, Vignali, & Dall'Olio, 2009).

URINE TESTING

Urine can be used for drug screening because urine is produced by the kidneys, which reabsorb and eliminate substances (such as drugs) that are waste products for the body (Cone, 1997). As a result, the substances that are being eliminated from the body will show up in the urine specimen. Most illicit drugs will be eliminated from the body within 48 hours of administration. If drugs are taken consistently and over longer periods of time, the detection time can be longer. The detection time also depends on the drug. For example, heroin, cocaine, and marijuana have a detection time of 1-3 days. Barbiturates, amphetamines, methadone, and methamphetamine have a detection time of 2-4 days (Cone, 1997). Thus, urine testing is very useful for the detection of short-term drug use, but it is not useful for assessing drug use that occurred more than a few days ago. This also means urine testing cannot be useful to examine drug use over long periods of time.

Another problem with urine testing is that the cut-off levels for detecting the drugs are considerably different depending on the drug, meaning that some drugs are detected at much lower doses than others. For instance, the cut-off level for marijuana is as low as 20 ng/ml, whereas the cut-off level for methamphetamine is 1000 ng/ml. Additionally, delay in drug testing (due to holding the specimen for a certain period of time before testing it) may also lead to a failure to detect drug use because the dose of the drug might have fallen below the cut-off level.

Despite these shortcomings, urine testing has been shown to be a valid measure of recent drug use. The advantages of urine tests have not only been demonstrated by social and behavioral research, but they have also long been recognized in the criminal justice system. For example, urine tests are used by courts to monitor abstinence and relapse, because offenders will very likely not report drug use to the

Court (Bureau of Justice Assistance, 1999). These studies and practices by the criminal justice system demonstrate the importance of using bioassays as a validation technique for self-reported drug use. In sum, the decision made by NIJ to opt for urine analysis as a validation method for recent self-reported drug use for the DUF and ADAM programs appears to be an adequate decision and was the major strength of DUF and ADAM over other national drug studies (such as the NSDUH and the MTF). The use of bioassays as a validation technique of self-reported drug use was not the only advantage of DUF and ADAM over the NSDUH and the MTF. Another important advantage was the collection of localized data.

LOCALIZED DATA COLLECTED EVERY QUARTER OVER 15 YEARS

As described earlier, the NSDUH, MTF, and DAWN provide national data on drug use only. Although they provide data for the major geographic areas (sub-state areas) of the United States, there is little data at the county and city levels. This county and city level data, however, is very useful for local law enforcement agencies and communities in their efforts to target hot spots of drug use and drug sale. DUF and ADAM filled that gap by providing county and city level data for the major drugs used. This site-specific information is important because the results of DUF and ADAM have consistently demonstrated that there are significant differences in drug use by geographic region and over time. For instance, the DUF program showed that methamphetamine use in the United States was modest, with about 6%, but it was becoming a great problem in the western part of the country (NIJ, 1996). Specifically, the six sites with the highest methamphetamine use rates were San Diego (37.1%), Phoenix (21.9%), Portland (18.7%), San Jose (18.5%), Omaha (8.1%), and Los Angeles (7.5%) (NIJ, 1996).

A study by the Drug Enforcement Agency (DEA) examining methamphetamine lab incidents across the United States shows that methamphetamine use has decreased in the western states and increased in central United States and eastern states (2011). Additionally, methamphetamine use increased over time, which showed especially in the number of drug treatment admissions and emergency room visits. Specifically, emergency room visits for methamphetamine related problems (provided by the Drug Abuse Warning Network) increased by 70% between 1999 and 2002 (Franco, 2007). Additionally, drug treatment admissions (provided by the Treatment Episode Data Set collected by SAMHSA) for methamphetamine use have also greatly increased, not only in the Western United States, but across the country. In 1992, only 5 states reported a high number of drug treatment admissions. In 2002, 21 states reported a high number of drug treatment admissions (NIDA, 2006).

The results from the MTF also demonstrate that methamphetamine was much more widespread in the western part of the United States. Unfortunately, questions about methamphetamine use were not included in the survey until 1999. Since 1999, the MTF suggests that methamphetamine use has been declining, from 8.2% in 1999 to 4.5% in 2005. The DUF data showed that methamphetamine began to rise sharply in the early 1990s. For instance, Herz (2000) using data from the DUF program for Omaha, Nebraska showed that methamphetamine use increased from 1% in 1990 to 10% in 1999. Being able to notice early when a certain drug is on the rise is important to prevent it from becoming an epidemic. Research suggests that recently implemented compstat systems can be used to track drug activity. These systems help local police departments determine areas with high crime and drug activity. For instance, the cities of Lowell and Newark were able to allocate police officers depending on the need of certain communities by using the compstat system (Willis, Mastrofski, and Weisburg, 2003). One problem with the local compstat systems is that they don't exist across the entire country. Also, the data collected by these systems is not send to a national data collection agency. As a result, these data are not available to researchers and others who are looking at trends across different geographic locations. The Drug Enforcement Agency uses methamphetamine lab incidents to track the prevalence of methamphetamine (2011).

Early knowledge about a rising drug problem enables law enforcement and the government to implement supply reduction strategies that reduce the availability of the drug on the streets and increase the prices for the drug. These federal and local prevention and intervention programs are only possible, however, if the problem is known not only to the local police department who uses a compstat system but also to departments in surrounding geographic locations. This is important because of a possible displacement of drug use and drug crime to other areas as a result of the increased police presence in areas with high drug use. As stated before, the distribution of the data is a necessary prerequisite to combating illicit drug use. DUF and ADAM were doing that by collecting data at the local level and making it available to researchers and policy makers.

This systematic data collection and distribution contributed to the implementation of local programs. In 1996, the government implemented the Comprehensive Methamphetamine Act, which aimed to reduce drug trafficking and reduce the availability of chemicals needed to produce methamphetamine. Researchers suggest that this Act was a direct response of the White House to the rising levels of methamphetamine use as documented by the DUF program (National Criminal Justice Association, 1999). Although the DUF program was not the only data that showed the regional increases of methamphetamine use, it provided important information that helped getting the attention of policy makers. For instance, Oregon reduced the availability of drugs (mainly pseudoephedrin) needed to produce methamphetamine (as provided by the Comprehensive Methamphetamine Act). Studies showed that during the time of supply disruption, methamphetamine use decreased substantially (Cunningham and Liu, 2003; 2005). The reduction of methamphetamine after the implementation of the Act was also evident in California and other states (National Criminal Justice Association, 1999). Without DUF and other programs, the sharp rise and extent of the methamphetamine problem might have gone unnoticed. As a result, the rising levels of methamphetamine use in the western part of the country might have further increased.

Similar to the MTF, NSDUH data also failed to show the substantial rise in methamphetamine use in the West. Specifically, the NSDUH show that methamphetamine use was less than 2% until 1994 and rose between 1994 and 2001 to 4%. Also, as described above, the MTF suggests that methamphetamine use has declined since 1990. In contrast, data from the ADAM program demonstrate that for some sites methamphetamine use has increased. Specifically, in San Jose, San Diego, and Phoenix, methamphetamine use increased substantially between 2000 and 2003. These results are also supported by the DAWN data, which shows an increase in emergency room visits for methamphetamine use between 1995 and 2002 in Los Angeles, Minneapolis, St. Louis, Seattle, Atlanta, New Orleans, and New York. Thus, the national data provided by the NSDUH and the MTF provide the false impression that methamphetamine is on the decline in general, when there are significant regional differences. Whereas methamphetamine use is declining in some regions, it is still increasing in others (Hunt, et al., 2006).

The revived ADAM II program, conducted by the ONDCP, also supports the finding that there are large regional differences in methamphetamine use and while some regions, such as Washington, DC and Portland, still show a decrease in methamphetamine use between 2003 and 2007, other sites (including Atlanta, Minneapolis, Sacramento, Indianapolis, Charlotte, and Chicago) have remained stable (Office of National Drug Control Policy, 2008).

Some police departments collect data on drug use in their community. For instance, the police department in High Point, North Carolina collects data about drug markets with the goal of targeting drug dealers more effectively (High Point Police Department). Although that is a viable approach if the goal is to target drug dealers and dismantle drug markets in a specific area, this does not necessarily advance research or help policy makers implement strategies for a greater region or nationwide because the data is not collected systematically. Instead, each local agency uses its own methods (e.g. surveys, GIS, crime reports) to determine hot spots of drug dealers and high drug market activity. This data is not comparable across agencies. Additionally, the local agencies do not submit their data to a national data bank where researchers could access such data. Systematic data collection at the national level ensures that researchers have data available to advance our knowledge about drug use and related issues. It is especially important to implement a national study that collects data systematically from arrestees because their drug using behaviors are substantially different from those of the general population and school children.

The NSDUH data not only failed to show the geographic differences for methamphetamine use but also greatly underestimated the extent of the methamphetamine problem. In fact, critics of the anti-methamphetamine drug policy argued that the NSDUH data does not support the degree of attention given to methamphetamine (Franco, 2007). This is problematic because research has shown that methamphetamine has stronger and longer lasting toxic effects than amphetamines or cocaine. For instance, smoking methamphetamine can create a high for 8 to 24 hours. In comparison, cocaine creates a high for only about 20 to 30 minutes. Additionally, the amount of dopamine released to the brain is three times higher for methamphetamine than for cocaine.

Also, it takes approximately 12 hours for half of the methamphetamine to be metabolized (half-life). The half-life of cocaine is only one hour. Methamphetamine has also been shown to increase the likelihood of HIV and Hepatitis infections due to risky sexual behavior. These findings indicate the great health consequences of methamphetamine use, and as a result the need to contain methamphetamine use when it first started to become more popular. In the 1990s, when neither the MTF nor the NSDUH provided valid information about methamphetamine use, DUF and ADAM were crucial to proving to critics that the implemented drug policies were very necessary to decrease the use of methamphetamine in the western part of the country and to hinder the further spread of methamphetamine use to other parts of the United States.

After having described the major advantages of the DUF and ADAM programs over the other national drug use monitoring programs the following section will now expand on differences and similarities between drug use data from the DUF and ADAM programs as compared to the NSDUH, MTF, and DAWN programs because these differences will demonstrate the importance of implementing a national program that systematically tracks drug use among arrestees.

CHANGES IN DRUG USE PREVALENCE AND PATTERNS BETWEEN 1988 AND 2002

Research from the DUF/ADAM programs, the NSDUH, and the MTF demonstrate that changes in drug use prevalence and patterns tend to show up among the criminal justice population first (DUF/ADAM data) and then spread to the general population (NSDUH and MTF data). With regard to marijuana use among arrestees, three waves (or generations) are visible between 1987 and 1998 (Golub and Johnson, 2001). First, marijuana use declined in the 1980s until about 1992. Second, beginning in 1992, marijuana use started to increase and then stabilized until about 1996. The third wave is characterized by a significant increase in marijuana use in 1996. This third wave appeared to plateau in 1999. Although these three waves are also apparent in the NSDUH and MTF, the beginning of a new wave showed up in the arrestee data from the DUF and ADAM data first and only later in the general population (Golub and Johnson, 2001).

Specifically, the increase in marijuana use became obvious first among youthful arrestees in the DUF data in 1991. This increase did not surface among the general population until about one or two years later (Golub and Johnson, 2001). The results further suggest that the rise in marijuana use was more pronounced among the criminal justice population, indicating that it spread more widely among this population as compared to the general population (Golub and Johnson, 2001). Additionally, the NSDUH shows a general decrease of marijuana use as individuals get older, the increase and decrease of marijuana use among arrestees was more dependent on geographic location. In some areas, marijuana use increased among older (born before 1967) arrestees (i.e., Los Angeles, Dallas, Denver, Houston), in other areas, marijuana use decreased among older arrestees (i.e., Portland, San Diego, San Jose) (Golub and Johnson, 2001).

There were also significant differences with regard to crack cocaine. Crack cocaine first appeared in the United States in the 1980s and quickly became popular (Golub and Johnson, 1997). Golub and Johnson (1997) examined the prevalence of crack cocaine between 1987 and 1996 using the DUF data. Their major findings demonstrate that crack cocaine became popular first among older, more experienced drug users, who then introduced crack to young and new users. Crack cocaine was cheap, easy to use (smoking), and widely available. Crack cocaine became an epidemic in the late 1980s and began to decline around 1996 mostly because younger drug users began to look down on crack users (also referred to as "crackheads") (Golub and Johnson, 1997). The rising disdain of young drug users for crack resulted in the decline of crack cocaine use. In contrast, the decrease in crack cocaine use among older, more experienced users was much less dramatic. In sum, the decline in crack cocaine use was not the result of a similar decline in the use of crack among all birth cohorts, but was caused mainly by the fact that youthful users stopped using crack cocaine (Golub and Johnson, 1997).

Whereas the DUF data showed that crack cocaine was being used widely between 1982 and 1996, the NSDUH did not distinguish between crack and powder cocaine until 1988. By that time, crack cocaine use was already decreasing, especially among young users. Thus, the NSDUH could not provide data about crack cocaine use in the general population at the time when crack was at its peak. Without the DUF and ADAM programs, researchers, law enforcement, and policy makers might not have known the true extent of the crack epidemic at the local and national levels. As stated above, even though local agencies might have had that information for their jurisdictions, this data is not available in a national database to researchers and policy makers.

The data also suggests that there was great geographic variation in the extent of crack use. The crack cocaine epidemic did not decline at the same rate across the country. Rather, the DUF data indicates that there was a significant decrease in crack cocaine use at 17 sites, still at its peak at five sites, and two sites did not have an epidemic. The popularity of crack cocaine varied even across sites that were geographically close. For instance, the decline of crack cocaine use in San Diego started in 1992 and was quite dramatic. Specifically, crack cocaine use declined from 37% in 1991 to 13% in 1996 among youthful offenders (23% decline). The decline of crack cocaine use in Los Angeles began around 1989, but was very slow. In Los Angeles between 1988 and 1996, crack cocaine use declined from 60% to 46% (14% decline). As was the case for marijuana and methamphetamine, the NSDUH data gives the impression that the rise and decline of crack cocaine happens at the same time across geographic locations in the country. As shown by DUF and ADAM, as well as DAWN, that is not correct. Again, this information is detrimental to the implementation of effective policies. There is no need for policies if crack cocaine is not being used. Similarly, the assumption that crack cocaine use is declining equally at all sites might lead to an abandonment of specific law enforcement strategies that might still be needed.

In sum, the findings of the DUF and ADAM programs demonstrate the differences in drug use prevalence and patterns found in the general population versus the population of arrestees. It also shows how important it is to monitor drug use among arrestees because trends tend to show up in the criminal justice population before they spread to the general population. DUF and ADAM were not only crucial for researchers examining drug using behaviors and the validity of self-reported drug use, but also to policy makers and law enforcement working to reduce drug use.

DRUG COURT MOVEMENT

The development of drug courts and diverting drug offenders into treatment services within the community has shown to be beneficial for the drug offenders and the community (ONDCP, 2011). Specifically, comprehensive drug treatment costs about $2,500 per year. In comparison, the costs of incarceration range anywhere between $20,000 and $50,000 per year for each person. Thus, drug treatment instead of incarceration can save states a sizable amount of money. Drug abusing offenders who receive treatment also have a significantly lower recidivism rate than drug abusing offenders who do not participate in drug court programs. The Office of National Drug Control Policy (2011) suggests that the recidivism rate of drug court graduates is less than 4%, compared to 66.7% of drug offenders released from prison (BJS, 2002).

DUF and ADAM provided information essential to the implementation of treatment and prevention programs for drug using offenders in communities by providing local data about drug using behaviors. As a result, it furthered efforts to lower recidivism rates among these drug abusing offenders. For instance, the ADAM data aided in the implementation of a new drug policy (Proposition 36) in Los Angeles County (Drug Use Alliance, 2006). Proposition 36 enables courts to place non-violent offenders with substance abuse issues in treatment programs rather than sending them to jail or prison. The DUF and ADAM data were important because they helped change the attitudes of policy makers and the public towards the treatment approach. The result of Proposition 36 is that in 2006 over 140,000 non-violent drug offenders received treatment (Drug Use Alliance, 2006).

This is important because the lifetime prevalence rates for substance abuse disorders among prisoners are between 68% and 74% (Karberg and James, 2005). The vast majority of these prisoners eventually return to the community, where they continue their drug using behaviors. Illicit drug use causes great costs for society because drug using prisoners have a high rate of recidivism and are very likely to engage in criminal activity. Specifically, Langan and Levin (2002) estimate that approximately two-thirds of drug involved offenders are re-arrested within three years of release from custody. These repeat offenders cause significant costs to the criminal justice system. The DUF and ADAM studies aided in the increase of funding for drug treatment in correctional facilities and encouraged the drug court movement. The drug court movement diverts drug abusing offenders away from the criminal justice system and provides them with drug treatment. There is considerable evidence that comprehensive drug treatment can reduce drug use and recidivism effectively. This is especially true when the drug treatment is followed by an aftercare program (Hora, Schma, and Rosenthal, 1998; Inciardi, Martin, & Butzin, 2004; Martin, Butzin, Saum, and Inciardi, 1999; Office of Justice Programs, 1998; ONDCP, 2011; Prendergast and Wexler, 2004).

DUF and ADAM were important for the progress made with regard to diverting drug offenders to drug courts for two reasons: (1) DUF and ADAM caught the attention of policy makers by showing the great extent of the drug use problem within the criminal justice population; and (2) by demonstrating the association between drug use and criminal behavior. For instance, the DUF findings indicate that increases in drug use and addiction lead to an increase in the crime rate (Fagan and Chin, 1990; Goldstein, 1990).

DRUG TREATMENT IN CORRECTIONAL FACILITIES

Additionally, correctional facilities have also expanded their treatment services. According to the Bureau of Justice Statistics (2004a), the percentage of jail inmates who received drug treatment increased from 39% in 1996 to 47% in 2002. Similarly, drug treatment programs have also increased for state and federal prisoners (BJS, 2002). DUF and ADAM provided crucial information to jail administrators implementing drug treatment programs.

DATA FOR LOCAL POLICE AGENCIES

Furthermore, DUF and ADAM provided information about drug use, drug markets, and hot spots for local police agencies. By providing local data about drug prevalence and patterns, specifically information about where arrestees buy their drugs and where they use drugs, DUF and ADAM helped local police agencies in their efforts to target drug sellers and users more effectively. Local police are a critical group in reducing the sale and use of illegal drugs. The ability to target specific areas known to be hot spots for drug sellers and users aids in the decrease of drug activity in those areas. This is also true for firearms.

DUF addenda collected data on firearm use which helped local police agencies. For instance, in Kansas City, police were able to target hot spots of firearm violence with gun seizures and significantly reduce the gun crime in these areas as a result (Decker, Pennell, and Caldwell, 1997).

DUF and ADAM were, however, important not only for law enforcement but also for policy makers and communities dealing with disease control. There is a strong connection between drug use and HIV and other infectious diseases. (e.g., hepatitis, tuberculosis) (Foundation of Drug Research, 2008). Among drug users, these infectious diseases are often spread via needle sharing and risky sexual behaviors (Zack, 2008). The Center for Disease Control and Prevention (CDC) found that 42% of AIDS cases stem from behaviors associated with drug use, either via needle sharing or unprotected sex. Additionally, 81% of AIDS cases in children are associated with transmission from the mother infected via drug injection and/or unprotected sex with a drug injecting person (CDC, 2004). Research has shown that substance abuse treatment is effective in reducing drug use and risky sexual behavior among offenders (World Health Organization, 2004).

For instance, the city of San Diego used information provided by the ADAM data to implement the Clean Syringe Exchange Program, which was found to reduce needle sharing and the spread of infectious diseases (Burke, 2004; City of San Diego, 2001). Similarly, research using the DUF data in combination with data from the Needle Exchange Program (NEP) has demonstrated that cities which implemented the NEP reduced needle sharing and drug injection, as a result decreasing the spread of HIV and other infectious diseases (DeSimone, 2005). This, in turn, reduces the costs for medical treatment and other services. The implications of these findings are that communities would benefit from educating drug users about the risks of unprotected sexual behaviors and needle sharing. Communities would also benefit from programs that proactively reduce risky behaviors, such as needle exchange (or similar) programs for drug users (Stephenson, et al., 2005).

EDUCATIONAL ATTAINMENT AND DRUG USE

Finally, data about drug use prevalence and patterns among arrestees is important because it demonstrates how crucial it is to keep kids in school and further educational attainment. There is a strong connection between educational attainment and drug use. As a result, there is also a connection between drug use and the number of unemployed people and people with low paying jobs. There is considerable evidence that people with low educational attainment are overrepresented among arrestees and especially among arrestees with substance abuse issues (Harder and Chilcoat, 2007). For example, DUF and ADAM data clearly demonstrate that the population of arrestees consists in great part of individuals with low educational attainment (i.e., no high school diploma), low income, and a lack of health insurance. This hinders their ability to seek treatment and other services that would increase their chances to break the cycle of drug use and offending.

Specifically, on average 30% of arrestees did not have a high school diploma, about 34% were unemployed, and a substantial proportion (15%) had no fixed address (NIJ, 2000). Research has also shown that a higher educational achievement is related to a better understanding of risky behaviors and the ability to find resources and treatment options that help address substance abuse issues and modify behaviors (Link and Phelan, 1995). These findings have implications for communities as they pertain to the importance of education in reducing drug use. Communities would benefit from programs that keep kids in school. Educational programs are certainly cheaper than the medical and criminal justice costs associated with drug offenders. These examples demonstrate the importance of systematically studying drug using behaviors among arrestees. Only if we understand these drug using behaviors can communities and law enforcement implement effective programs that aim to reduce drug use and provide services to individuals who are using drugs.

LACK OF NATIONAL STUDY MONITORING DRUG USE AMONG ARRESTEES

To reiterate, currently there is no national study examining drug use prevalence and patterns among arrestees because the revived ADAM II program (conducted by the ONDCP) only collects data for 10 counties within the United States. To reiterate, even though local agencies might be collecting similar drug use data, this is not a systematic data collection and therefore results in a number of problems associated with the use of such data. These problems are: (1) it makes it more difficult for researchers to obtain such data, (2) it is not publicized widely where such data exists, and (3) methodological differences between local data collection might make it difficult to compare data across sites.

Researchers argue that the money spent on examining drug using behaviors is distributed improperly. The National Household Survey costs $50 million a year, but it examines a population group that is at low risk of using drugs. Kleiman (2004) stated that for cocaine, the NSDUH accounts for about 10% of the actual cocaine consumption, leaving the other 90% unexplained.

Additionally, the NSDUH also does not examine the revenues of illicit drug markets or the crimes associated with drug use. The amount of drugs consumed, illicit drug markets, and the relationship of crime with drug use is arguably more important than simply knowing the number of people who has used drugs in the last year (Kleiman, 2004). Drug markets and crime constitute a significant problem for society and are one of the top priorities of the government. Research has demonstrated that, in many cases, criminal behavior (e.g., property crime, prostitution, drug sales) results directly from attempts to support a drug habit. For instance, Corman and Mocan (2000), in their time-analysis study in New York City, found that robberies, burglaries, and motor vehicle theft increased during the time period of increased drug use. Additionally, according to a survey by the Bureau of Justice Statistics (2004b) of federal and state prison inmates, an estimated 17% of state prisoners and 18% of federal prisoners reported committing offenses in order to support their drug habit.

The next chapter will introduce the DUF program, examine the criticisms brought forth by the GAO and analyze the changes made to the new ADAM program.

CHAPTER 3

DUF and ADAM Program: Criticism and Transition from a The Convenience to a Probability Sample

UNDERLYING ASSUMPTION OF THE DUF STUDY

The DUF study was developed under the assumption that drug users are likely to be among the population of arrestees (BJS, 1998, Mallender, Roberts, and Seddon, 2002). A number of researchers have supported that assumption (Brook, Whiteman, Finch, and Cohen, 1996, Wish and Gropper, 1990), and several explanations have been advanced regarding this finding. First, drug users are more likely to be involved in criminal behavior because they need a certain supply of money to pay for the drugs and have to use illegal means (such as burglary and robbery) to obtain such monies (Petersilia, Greenwood, and Lavin, 1978, Goldstein, 1987). Second, drug users may be more likely to engage in violent behavior due to emotional and/or mental reactions such as aggressiveness and irritability caused by certain drugs (Bickel and DeGrandpre, 1996). Third, drug users are more likely to deal drugs to support their drug habit and more likely to get into "turf wars'" amongst drug dealers and as a result get arrested (Goldstein, 1987). There is also evidence, however, that the relationship between violent crime and drug use may not be as strong as assumed and that higher arrest rates of drug users for violent crime may instead be due to drug enforcement policies and procedures (Resignato, 2000). Said differently, the number of drug users who commit violent crimes may not be significantly different from the number of violent crimes committed in the whole population, but because police resources focus on drug users/drug dealers, they are overrepresented among arrestees.

Regardless of the reasons, the high likelihood of finding a larger number of drug users among arrestees makes prison and jail populations the most suitable sampling frame when studying drug use, changes in drug patterns, and characteristics of drug users, because arrestees were seen as the leading indicator as compared to other population groups. Stated differently, the sampling frame was based on research showing that new drug use patterns and drug paraphernalia show up first among arrestees and then spread to other population groups. With this approach, the DUF study filled an important gap in the research on drug use prevalence and patterns because other drug studies (e.g., the National Household Survey and Monitoring the Future) studied populations such as households and school children, among which drug use is much less widespread. Additionally, no other national drug study besides DUF used bioassays as a validation technique for the self-reported drug use, making DUF an incredibly valuable tool for researchers studying not only prevalence and patterns of drug use but also the validity of self-reported drug use.

DATA COLLECTION IN THE DUF STUDY

The quarterly data collection was based on voluntary participation of arrestees for bot the self-report interview and the urine test. The data collection began in 1987 in 12 cities across the United States. DUF increased the number of sites steadily over time from 12 sites in 1987 to 21 sites in 1988, 22 sites in 1989 and 24 sites for the years 1990 until 1997 (National Institute of Justice, 1998). The 24 sites were Atlanta, Birmingham, Chicago, Cleveland, Dallas, Denver, Detroit, Ft. Lauderdale, Houston, Indianapolis, Kansas City, Los Angeles, New York (Manhattan), Miami, New Orleans, Omaha, Philadelphia, Phoenix, Portland, St. Louis, San Antonio, San Diego, San Jose, and Washington, DC. The DUF sites were not distributed evenly across the United States, however. Rather, the majority of sites were concentrated in the Pacific region, the southern part of United States, the middle of the United States, and the East Coast. California (4), Texas (3), and Florida (2) had multiple sites (National Institute of Justice, 1998).

NUMBER OF ARRESTEES

The number of booked male arrestees varied from year to year. As can be expected, with the increasing number of sites, the number of male arrestees also increased. The increase was, however, not consistent. The number of arrestees peaked in 1991 with 22,335 male arrestees and then declined steadily to 19,736 male arrestees in 1997. For an easier overview, Table 3.1 shows the number of male arrestees and number of sites for each DUF year (National Institute of Justice, 1998).

Table 3.1. Number of Male Arrestees and Sites by Year: DUF

Years	Male Arrestees	Number of Sites
1987	2,993	11
1988	10,548	20
1989	16,186	21
1990	20,556	23
1991	22,335	24
1992	22,265	24
1993	20,551	23
1994	19,987	23
1995	20,737	23
1996	19,835	23
1997	19,736	23
1998	20,715	35

The table shows that the number of interviewed arrestees increased simultaneously to the increasing number of sites between 1987 and 1991. Within that time period, the number of interviewed arrestees increased from 2,993 arrestees and 23 sites during the first year of the program to 22,556 arrestees and 24 sites in 1991 and 22,265 arrestees and 24 sites in 1992. In 1993, the sample size declined somewhat, probably due to the reduction of sites from 24 to 23. In the following years, the number of interviewed arrestees remained relatively stable and the number of sites was 23 until 1998. Even though 1998 is considered by NIJ to be the first ADAM year, the probability sampling procedure was not implemented until the second half of the year 1999.

For the purpose of this research, the year 1998 will be considered a DUF year because the sampling method was the same as in the previous years.

DESCRIPTION OF VARIABLES

The collected data includes information for the following five topics: (1) drug use by type of drug and for each individual offender (self-report and urine analysis); (2) dependency on alcohol/drugs (self-report); (3) need for treatment (self-report), (4) relationship between drug use and crime (offenses); and (5) indicators of self reported drug use compared to indicators of drug use according to the urine analysis (National Institute of Justice, 1997).

Arrest records were used to obtain information about birth year, race, and the top charge. The DUF questionnaire included items regarding participant demographic characteristics - namely gender, race, marital status, educational attainment, employment status, and living circumstances (National Institute of Justice, 1998), and lifetime and recent drug use (within the past three days) of 22 drugs for the years 1987 till 1995 and 15 drugs for the years 1996 and 1997.

For each drug, arrestees were questioned regarding: (1) age of first use; (2) frequency of use during the past month; (3) recent drug use (past three days); (4) route of administration; (5) perception of past dependency; (6) perception of current dependency; and (7) past drug treatment. Additionally, arrestees were questioned regarding arrests during the past 12 months and whether they were under the influence of drugs at the time of the crime. Besides questions about demographics, drug use, and arrests, the questionnaire contained items regarding how much money arrestees spent on drugs during an average week and whether they had been in the emergency room for drug-related incidents (National Institute of Justice, 1998).

Additionally, NIJ used several addenda to assess topics of interest that were not typically part of the survey. For example, in 1995, the survey included a heroin addendum and in 1996 a heroin and gun addendum. Another addendum asked arrestees about their knowledge and consideration of AIDS when using intravenous drugs (National Institute of Justice, 1997). These addenda were not collected systematically; rather, they were only collected at certain sites for a limited period of time (National Institute of Justice, 1997).

RESPONSE RATE

The datasets provided by NIJ only include data on arrestees that agreed to the interview and completed both the self-report interview and the urine analysis. Approximately 90% of those asked to participate agreed to do so and of those, 80% provided a urine sample (National Institute of Justice, 1997). There is, however, no information about arrestees who were either not asked to participate or did not consent to an interview.

DRUG TESTING

The urine analysis included screening for 10 drugs (marijuana, opiates, cocaine, barbiturates, amphetamines, PCP, methadone, benzodiazepines (Valium), methaqualone, and propoxphene (Darvon) via radioimmunoassay (National Institute of Justice, 1995). To ensure that positive urine tests for amphetamines were correct, gas chromatography was used. The urine tests for all DUF sites were done at a central location. The outcome of all urine tests was dichotomous, that is either positive or negative for the tested drug (NIJ, 1995).

DUF SAMPLING PROCEDURES

The sampling design was a non-probability sample guided by target numbers of interviews (250 males and 100 females) and a priority charge system. The data was collected over a period of 10 days or until the target sample number had been reached (McBride and Swartz, 1990). The problems associated with this type of sampling will be explained in greater detail in the next section when discussing the GAO report and the criticisms brought forth by the GAO.

The process of obtaining participants involved a number of steps. First, arriving arrestees were brought into the booking area of the facility (holding cell, central booking area, or other applicable area); second, the site administrator explained the study and asked for volunteers. The site administrator told the arrestees only about the self-report interviews. At that time, the arrestees did not know that they would also be asked to provide a urine sample. As an incentive to participate, most sites offered candies, cigarettes, or coffee. Third, the site administrator recorded the top charges for each arrestee and recruited participants based on the priority charge system. The site administrator attempted to recruit volunteers with non-drug felony charges, followed by non-drug misdemeanor charges, and finally offenders with drug-related charges. Fourth, each participant received an ID that matched the survey and the urine test. Fifth, the participant completed the interview. After the completion of the survey, the participant was asked to provide a urine sample. If the arrestee provided the urine sample, a staff member would then collect the sample and label it with the same ID as the questionnaire (Swartz, 1990).

All self-report interviews and the urine test results were gathered after each quarter by NIJ. NIJ checked the accuracy of the data by looking at the consistency of answers and undocumented codes. The NIJ staff also standardized the missing data codes across all sites. The self-report data was then merged with the urine test data and the complete dataset was re-formatted to make it usable for researchers. Finally, the datasets were made available to researchers via the Inter-University Consortium for Political and Social Research (www.icpsr.umich.edu) (National Institute of Justice, 1997). The following section will discuss the criticisms brought forth by the GAO and how these criticisms led to a change from a non-probability sample to a probability sample implemented in 1998.

GAO REPORT CRITICISMS

Although the GAO report (1993) highlighted the strengths and unique features of the DUF program, especially the collection of drug use data from arrestees in different cities across the country and the validation of the self-report data via urine analysis, the report also heavily criticized the non-probability sample used by DUF, stating that due to its non-probability sample and lack of standardization across sites, the sample of arrestees in the DUF study may not provide accurate information about arrestees in general and thus results obtained from the study about drug use prevalence and patterns may not be accurate (GAO, 1993). The GAO report criticized DUF for three major reasons, all of which pertained to the sampling design: (1) the selection of booking facilities included in the study; (2) the subject sampling procedure; and, (3) the inclusion and exclusion criteria in selecting arrestees within the booking facilities for the interview.

(1) Criticism on the Selection of Booking Facilities

The DUF program collected self-report data and urine tests at central booking facilities in a number of cities across the country. The GAO report criticized that the different booking facilities represented very different geographic units including entire cities or even counties, parts of a city or a county, or a central city plus additional cities. The GAO report took issue with the fact that booking facilities encompassed very different geographic areas, as a result making it difficult to draw conclusions about drug use prevalence and patterns for the greater area. The GAO report stated that the limitations of the study due to the selection of booking facilities hindered the development of drug policy and programs: "There is no evidence to support generalizing partial data to an entire city or county" and "Caution is warranted in using these data to determine booked arrestee drug prevalence rates" (GAO, 1993, p. 52). This statement leads us to the second major criticism regarding the sampling design: the subject sampling procedure.

(2) Criticisms of the Subject Sampling Procedure

The DUF program used a judgment-based sample of arrestees based on a target number of interviews (225 male and 100 female arrestees per quarter per facility). Starting in 1990, the sampling of the subjects was based on the 20% rule, meaning that every fifth arrestee interviewed should have been charged with a drug offense (GAO, 1993). The GAO pointed out that the 20% rule lead to unpredictable consequences. It can either lead to an underestimation of the drug use prevalence among arrestees if more than 20% of the arrestee population uses drugs, or it can lead to an overestimation of the drug use prevalence among arrestees if less than 20% of the arrestees use drugs (GAO, 1993). The criticisms pertaining to the target rate of offenders and the 20% rule are closely related to the inclusion and exclusion criteria of arrestees in the DUF study.

(3) Criticism of the Inclusion and Exclusion Criteria

The third major sampling issue highlighted by the GAO report pertains to the criteria used to select arrestees—the inclusion and exclusion criteria. The criticisms by the GAO focused on the fact that there were no standardized procedures across sites. Each DUF site used its own criteria within some set standards. For example, the DUF program established a rank order of criminal charges that was used to select male arrestees. Each DUF site, however, made its own decisions about which types of offenders would be interviewed. For instance, the San Diego site eliminated all misdemeanor offenders, and the Miami site only interviewed male offenders arrested on felony charges. Additionally, in New York (Manhattan) the booking facility limited the number of misdemeanor offenders, as a result decreasing the number of misdemeanor offenders available for interviewing. In Omaha, Nebraska, all male arrestees were interviewed (disregarding the charge rank-order altogether) due to the small arrestee pool available. Six sites (Denver, Detroit, Fort Lauderdale, Kansas City, Houston, and Indianapolis) had access to arrestees who committed criminal offenses in court, in jail, or in custody. The other sites either eliminated this offender group or simply did not have access to them. For female arrestees, there were no rank orders or other guidelines besides the target interview number of 100 females per quarter per site (GAO, 1993).

In sum, each DUF site worked differently, and the implications of these differences were heavily criticized by the GAO concluding: "Evaluators of the data are thereby unable to determine whether decreasing or increasing drug use scores represent statistically significant shifts in actual drug use. An individual's conclusion about drug use patterns and trends must therefore rely on intuitive reactions rather than being statistically based" (GAO, 1993, p.57).

The GAO report and the criticisms brought forth therein had great consequences for the study. In 1996, NIJ began restructuring the methodology of the drug use forecasting study and it was re-named the Arrestee Drug Abuse Monitoring Program (ADAM). The purpose of the restructuring was to enable the researchers to use inferential statistics and obtain results that could be generalized to the larger population. In 1998, NIJ implemented the ADAM study. With the implementation of the new study came a number of changes in the data collection process, as well as in the questionnaire itself.

OVERVIEW OF THRE ADAM STUDY

The redesign of DUF/ADAM began in 1997 and consisted of three components: (1) expansion of the program; (2) implementation of a probability sample; and (3) implementation of a redesigned data collection instrument (Yacoubian, 2004). The goal of these changes was to produce results that could be generalized to the target population of booked arrestees and to make the program more valuable and useful for practitioners and policy makers (NIJ, 1998). The first changes to the study were made in 1997 and 1998 by expanding the number of sites from 23 to 35 and by standardizing the sampling procedure across sites (Yacoubian, 2004). NIJ began implementing the redesigned study in the second half of 1999, but it was not fully implemented until the beginning of the year 2000 (Arrestee Drug Abuse Monitoring (ADAM)). Next, the new sampling procedure will be described, followed by a description of the expansion of the program and the new data collection instrument.

Starting in 1999/2000, NIJ implemented complex data collection procedures that would ensure standardized outcomes. The study attempted to collect data on offenses, offenders, and drug use that would be a representative mix of the larger population. To achieve this goal, NIJ applied the same definitions of "catchment areas" to all sites. "Catchment areas" are "regions from which arrestees are drawn at the sites" (NIJ 2000, p. 178). The definition of these "catchment areas" are: (1) persons taken to a booking facility, and (2) the "catchment area" is a county.

In addition, NIJ standardized the definitions of offenses because of the wide variety of definitions for certain offenses among counties and states. First, in order to ensure that all arrested persons had the same probability of being sampled, only "booked arrestees" were included in the study. This specification was necessary to avoid a bias in the sample towards more serious offenses, because not all counties handle arrests the same way. Some counties arrest and book all persons; others arrest but then release the persons if the offense committed was a misdemeanor. Second, NIJ standardized the definitions for terms such as misdemeanors and felonies vary. What is defined as a misdemeanor offense in some counties is considered a felony offense in other counties, depending on state law. Some counties distinguish misdemeanors from felonies in regards to the length of jail time; other counties distinguish misdemeanors from felonies based on the seriousness of the crime (NIJ 2000).

NIJ also standardized training procedures for all sites to guarantee that the data collection procedures were not only the same at each site but also fully followed at each site. NIJ staff monitored the training as well as the data collection by implementing performance standards and feedback to each site after every data collection cycle. In addition, follow-up training sessions were conducted at regular time intervals to ensure that the interviewers followed the procedures strictly and consistently (NIJ 2000).

SAMPLING PROCEDURES 1998

In 1998, ADAM expanded the number of sites considerably from DUF (24 sites) to a total of 35 sites for adult male arrestees, 32 sites for adult female arrestees, 13 sites for juvenile male arrestees, and 8 sites for juvenile female arrestees where interviews were completed and urine specimen collected. Interviews were conducted with arrestees who had been at the facility for no more than 48 hours. In 1998, NIJ responded to one of the criticisms of the GAO by standardizing the catchment areas. The catchment area was now the same for all sites—the county (NIJ, 1998).

The target number of interviews plus urine specimens was 225 for adult male arrestees (all 35 sites), 100 for adult female arrestees (32 sites), 100 for juvenile male arrestees (13 sites), and 100 for juvenile female arrestees (8 sites) per quarter. However, the sampling was still conducted on a voluntary basis as a judgment-based sample.

SAMPLING PRCEDURES 1999

In the third and fourth quarter of 1999, NIJ began for the first time to employ a probability sample. The sampling plan is very similar to the sampling plan for the years 2000 to 2003 (explained in detail below). The differences are: (1) in 1999, interview shifts lasted between 4 and 8 hours, whereas interview shifts lasted 8 hours starting in 2000 and (2) in 1999, the sampling, data collection, and training procedures were still in process of being fully implemented.

SAMPLING PROCEDURES 2000 – 2003

Beginning in 2000, NIJ fully implemented the changes to the design of DUF/ADAM. The sampling plan included a probability-based sampling design for all persons arrested and booked at each site. The sampling method was tailored to each county and its characteristics. In addition, the sampling strategy itself was tailored to each site and the sample was weighted. The goal behind this new sampling design was to "represent with known probability the likelihood that a male arrestee was selected for an interview and to use that information to weight each sample case" (NIJ, 1998, p. 7). Another change to the design was that now every day of the week and every hour of the day were represented in the sample to ensure that no bias would occur due to the possibility that interviewers were not present at the time the person was arrested and booked. Hence, all persons arrested within the last 48 hours of the beginning of the work shift were included in the population from which the sample was drawn. In addition, the new design included booking facilities representing all types of facilities from small to large, urban and rural, quick release and slow release to avoid a bias and include all types of offenders (NIJ, 2000).

THE SITE'S SAMPLING DESIGN

NIJ employed one of four sampling models depending on the characteristics of the county. A total of four sampling models were necessary to ensure that the sampled persons would be representative of the population of the whole county. First, the "single jail" sampling model was developed specifically for counties where data was collected only in one jail. Second, for counties in which there were between two and six jails, a "stratified" sampling model was employed. The "stratified" model allowed for data collection in all jails by determining a target number of arrestees to be sampled. In each jail the target number of arrestees to be sampled was proportionate to the number of arrestees booked. Third, a "stratified cluster" model was utilized in counties with more than six jails. Each jail was included in the strata. From each stratum one or two jails were sampled. Fourth, NIJ developed a "feeder" model from the stratified cluster sample for counties with booking facilities which transferred booked arrestees very quickly to a central holding facility. The "feeder model" became necessary in those counties because the booked arrestees would have been inaccessible for interviews. In counties where the "feeder" model was utilized, booked arrestees were interviewed at the holding facilities and in the jails that transferred arrestees to those holding facilities (NIJ, 2000).

WEIGHTING OROCEDURE

In addition, the data was weighted to guarantee that the sample was representative and generalizable to the county population. ADAM employed a non-traditional method to determine the weights, because at the time of the sampling it was unknown who would be arrested. Hence, the probability with which one were to be sampled for the interview was also unknown. The weighting of the data was based on the following assumptions: (1) "Arrestees charged with more serious crimes spend more time in jail facilities than those arrested on less serious charges;" (2) "Arrestees booked at the same time of day are processed similarly; that is, they all spend approximately the same amount of time in the jail before arraignment and/or transfer to another holing facility,;" (3) "The stock and flow model may mean more serious offenders will be over-represented in the stock population, while the flow sample should represent all charges." [The assumption was that by using this procedure all arrestees should have a similar chance of being selected into the sample]; and, (4) Arrestees who are booked on days where many arrestees enter the facility had a lower chance of being selected for the sample than if they were arrested on days when few arrestees enter the facility (NIJ, 2000, p. 181).

A "post-sampling stratification" design was developed to calculate the "probability of inclusion in the sample of like groups of arrestees" (NIJ, 2000, p. 181). NIJ collected census data for the total population of arrested individuals for each collection site. Each collection site also provided demographic, booking, and offense information about all persons arrested and booked at the site where the interviews took place. The information obtained from each site was then compared to the sample collected at each site and match with the county population. Furthermore, based on the information about the characteristics of arrestees and the county population, the population was divided and strata developed. Each stratum contained a certain number of arrestees and from this information the probability of each arrestee to be sampled compared to all other arrestees was calculated (NIJ, 2000).

The weighing process can be affected by three problems: (1) *Ineligibles*, defined as persons that are not eligible to be part of the study, such as persons on extradition, court, and federal holds; (2) *Duplicates*, defined as persons that are entered into the system more than once because of aliases or mistakes of the staff; and, (3) *Inconsistent booking times* because jails vary on their definition of booking times (intake v. time of intake) and the time at which the booking time is logged by the jail (NIJ, 2000, p. 181).

THE FACILITY-LEVEL SAMPLING DESIGN

The sample size for each facility was proportionate to the number of persons arrested and booked in each county. The arrest and booking information was provided either by the site staff of the facility (if possible) or by the FBI's Uniform Crime Reports (if the information was not available from the site staff). The number of booked arrestees to be sampled was proportionate to the number of facilities in the county and the number of persons booked at each facility. The target number of interviews (per quarter) depended on the sampling design of each facility and the number of booked arrestees in each facility. Interviews were conducted every day of the week for one-to-two weeks during the 8-hour period in which the intake was at its highest. The number of interviewers remained constant throughout the work shift (NIJ, 2000, p. 181).

As stated above, the population to be sampled must include all persons arrested and booked, at all days of the week and all hours of the day. To facilitate such a sampling design, each day was split into "Stock" and "Flow" periods. The "stock" period is the part of the 16 hours per day in which interviewers are not present. In order to sample persons from the "stock," the jail staff provided a list of all persons booked during those 16 hours of the day. From that list, which is ordered by the booking time, booked arrestees were sampled depending on the target number in even intervals. The "stock" interviewers stopped working once they reached the target number. Once determined, the number of interviewers did not change at the sites. This was mostly done due to practical reasons to make the implementation of the new sampling procedures easier and to avoid problems with the weighting procedure (NIJ, 2000).

The "Flow" period represented the part of the day during which arrestees were interviewed, which was eight hours each day. The "Flow" hours were tailored to each site, with the goal to have interviewers present during the 8 hours of the day in which the number of arrested and booked persons was the highest. Persons were sampled from each period, the "Flow" and the "Stock." Flow interviewers work continuously during these 8 hours regardless of whether the target number of interviews for the day has been reached. After each interview, the ADAM staff sampled the persons booked closest to the time of the completion of the interview (NIJ, 2000, p. 181). The sampling was carried out for a period of one to two weeks at the time (NIJ, 2000).

FACE SHEETS

The ADAM staff filled out a face-sheet (information about booking and charges) for each arrestee sampled, regardless of whether the arrestee agreed to interview and regardless of whether the arrestee was available for an interview. This procedure decreases potential bias that could occur and enables the site staff to check whether the correct sampling procedure has been followed (NIJ, 2000). In facilities where arrestees are transferred to other facilities very quickly or released very quickly it sometimes became impossible for the ADAM staff to meet the quota for the stock interviews. In such cases, it was possible to have separate interview periods for stock and flow (NIJ, 2000).

RESPONSE RATE

The response rate for male arrestees in 2000 was 56.3%. Of the non-respondents, 13.6% declined to interview, 22.8% were not available (either released, in a holding cell, or in medical unit), and 7.3% were not asked to participate. Of those 56.3% that agreed to interview, 89.9% also provided a urine sample. Similar to the year 2000, the overall response rate for male arrestees in 2001 was 55.1%. Of the arrestees that did not respond, 12.2% refused to interview, 22.5% were not available (either released, in a holding cell, or in medical unit), and 10.2% were not asked to participate for various reasons. Of those who interviewed, about 90.5% percent also provided a urine sample. It is difficult to compare these response rates to that of DUF because DUF used a convenience sample and therefore contains no information about non-respondents. The DUF documentation also does not include information on how exactly the response rate was calculated.

Arrestees who were available for the interview might differ in their drug using behaviors from arrestees who were available for the interview. For instance, arrestees who had already been released were more likely to have committed misdemeanor offenses and/or were less likely to have used drugs. Similarly, arrestees who were dangerous or violent might also differ in their drug using behaviors as compared to arrestees who were compliant with law enforcement. There exists no data on these unavailable arrestees that would allow for a comparison. Although this is speculative, it is possible that the probability sample of ADAM might have included arrestees very similar to the arrestees in the DUF program - that is, arrestees available at the time of the interview and who were willing to participate.

The ADAM data also suggests that for both years, 2000 and 2001, the response rate of arrestees who were asked to participate in the ADAM study varied greatly depending on the site. In 2000, there were 35 sites at which data was collected for at least one quarter. Half of these sites had a response rate of 81% or higher for arrestees who were available and asked to participate. Overall, the response rate varied between 5.9% and 40.1%, with Fort Lauderdale being the lowest and the Charlotte Metro area being the highest (NIJ, 2003). Similarly, the percentage of interviewees who agreed to the urine test also varied considerably between 74.7% and 96.6%, with the Albany/Capital Area having the lowest rate and Fort Lauderdale having the highest rate (NIJ, 2003). These response rates were lower as compared to the DUF data, which had a 90% response rate for interviews. This is about 9% higher than the average response rate for the ADAM program (U.S. Department of Justice, 1998). Again, for the DUF data there is no data that would allow for a comparison of the arrestees who refused to the arrestees who participated. The differences in the response rates for DUF and ADAM can likely be attributed to the differences in the sampling method. During the DUF study, the local DUF interviewers had some discretion in whom to ask to volunteer. The sample was not a probability sample of all booked arrestees, rather the interviewers would ask for volunteers (based on certain criteria described earlier). Additionally, many sites offered incentives to the arrestee to increase their willingness to participate. This might explain the higher response rates for DUF as compared to ADAM.

Contrary to DUF, the ADAM program included a face sheet for all arrestees selected for the sample. The face-sheet included information about gender, race, age, residence, charge type, and charge seriousness. A study by Myrstol and Langworthy (2005) used this face-sheet data to compare the realized sample of arrestees in Anchorage with the arrestees who were not included in the sample. The authors showed that despite a relatively high attrition rate, both the male and female realized sample was representative of the population of booked arrestees.

DRUG TESTING

Table 3.2 shows cut-off levels and detection period for the drugs examined in the ADAM study. Contrary to DUF, all ADAM sites sent the urine specimen to a central facility where it was tested for 10 drugs, as a result standardizing the drug testing. The table includes cut-off levels and detection periods for each drug. The cut-off level is defined as the "amount of the drug in nanograms per milliliter below which the amount is considered undetectable and the result is negative" (NIJ, 2003, p. 16). The detection period is defined as the "number of days after ingestion during which the drug can be detected in the body" (NIJ, 2003, p. 16).

Table 3.2. Cut-off Levels and Detection Periods for 10 Drugs

Drug	Cut-off Level	Detection Period
Cocaine	300ng/ml	2-3 days
Marijuana	50ng/ml	7 days (infrequent use) 30 days maximum (chronic use)
Methamphetamine	300ng/ml	2-4 days
Opiates	300ng/ml	2-3 days
PCP	25ng/ml	3-8 days
Amphetamines	1,000ng/ml	2-4 days
Barbiturates	300ng/ml	3 days
Benzodiazepines	300ng/ml	2 weeks maximum
Methadone	300ng/ml	2-4 days
Methaqualone	300ng/ml	10 days maximum
Propoxyphene	300ng/ml	3-7 days

SITES

The ADAM study started with a considerable increase in the number of sites and states included in the study compared to DUF. ADAM started out with 35 sites in 1998, 12 sites more than the DUF study had in 1997. ADAM then increased the number of sites to 39 sites in 2001. The 39 sites were Albany/Capital Area (NY), Albuquerque (NM), Anchorage (AK), Atlanta (GA), Birmingham (AL), Charlotte-Metro Area (NC), Chicago (IL), Cleveland (OH), Dallas (TX), Denver (CO), Des Moines (IA), Detroit (MI), Ft. Lauderdale (FL), Honolulu (HI), Houston (TX), Indianapolis (IN), Kansas City (MO), Laredo (TX), Las

Vegas (NV), Los Angeles (CA), New York (NY), Miami (FL), Minneapolis (MN), New Orleans (LA), Oklahoma City (OK), Omaha (NE), Philadelphia (PA), Phoenix (AZ), Portland (OR), Sacramento (CA), Salt Lake City (UT), San Antonio (TX), San Diego (CA), San Jose (CA), Seattle (WA), Spokane (WA), St. Louis (MO), Tucson (AZ), and Washington, DC.

Although NIJ added a considerable number of sites (14) to the ADAM study, the distribution patterns is similar to the distribution of sites in the DUF study. The majority of sites were concentrated in the Pacific region, the southern part of United States, the middle of the United States, and the East Coast. California (4), Texas (4), Florida (2), Missouri (2), and Arizona (2) had multiple sites. The ADAM sites were distributed across 26 states.

NUMBER OF ARRESTEES

The number of arrestees increased steadily between 1998 and 2001. Table 3.3 represents the number of male arrestees for each year congruent with the increasing number of sites where data was collected. Between 1999 and 2001, the number of interviewed arrestees steadily increased without an increase in the number of sites. By 2001, the sample size was nearly twice that of the last DUF years.

Table 3.3. Number of Interviewed Arrestees and Number of Sites by Year

Year	Male Arrestees	Number of Sites
1999	31,210	35
2000	35,784	35
2001	39,406	39

DESCRIPTION OF VARIABLES – REDESIGNED DATA COLLECTION INSTRUMENT

Similar to DUF, arrest records were used to obtain information about race, gender, birth year, and top charge (NIJ, 2003). Also similar to DUF, ADAM collected information on demographic characteristics, recent and long term drug use, and drug use patterns (NIJ, 2003).

There were, however, a number of modifications to the ADAM questionnaire. The ADAM questionnaire was expanded to include screening questions for drug dependence and need of treatment, information about drug use within the previous year via the calendar method, drug market activity, and drug treatment by drug within the past year. The drug dependence and need of treatment screening consisted of questions regarding the frequency of drug use, thoughts about drug use, reasons of drug use, intentions to reduce or stop drug use, and objections from friends or family to the drug use (NIJ, 2003).

CALENDAR METHOD

The implementation of the calendar method was probably the most important change in the questionnaire (NIJ, 2003). The main purpose of the calendaring method was to collect drug use information within the past year and to improve the validity of self-reported drug use among arrestees. The basic idea was to examine annual patterns of drug use and related behaviors over time because asking arrestees about drug use only during the last 30 days does not capture the complexity of drug use patterns (NIJ, 2003). The method of calendaring was utilized to help the respondents remember drug using behaviors over such a long period of time. To help arrestees remember what drugs they used and how frequently they used them, interviewers first asked arrestees about major events during the last year (e.g., birthday, holidays, family events, and life events) with the purpose of conceptually dividing the year into manageable time periods for the respondents. The major events reported by the arrestee serve as an "anchor" helping them remember their drug using behavior at that time. Using this method, data on drug use patterns was collected for each drug month by month (NIJ, 2003).

As stated above, one of the main goals of the calendaring method was to improve the validity of self-reported drug use. Whether this goal was accomplished has not been extensively assessed. A study by Yacubian (2003b) concluded that the calendaring method did not influence the validity of self-reported drug use. The results demonstrated that the concordance rate between self-reported drug use and the urine analysis tests for the ADAM years during which the calendaring method was used (2000) were similar to the concordance rate for those years during which the calendaring method was not used (1999). Specifically, for the year 1999 (no calendaring) the concordance rate was 64% for marijuana, 48% for cocaine, and 8% for heroin. Similarly, in 2000 (calendaring was used) the concordance rate was 59% for marijuana, 47% for cocaine, and 6% for heroin. Thus, the calendaring method did not improve the validity of self-reported drug use.

The modified ADAM questionnaire also included a section on drug market activity including questions about drug purchase patterns, place and neighborhood of purchase, and difficulties purchasing drugs (NIJ, 2003). More specifically, arrestees were asked a battery of questions about how often they bought drugs, how much they bought, who they bought drugs from, how they contacted the drug dealer, the relationship of buyers and sellers, how much they paid for the drugs, the payment method, the neighborhood in which they bought the drugs, how they located drugs, difficulties in locating and buying drugs, and why drug purchases failed. This information was collected for each drug (NIJ, 2003). How accurate this information is unknown. It could be expected, however, that recall decay and forward telescoping might factor into this issue.

DRUG TREATMENT DATA

Finally, the modified ADAM questionnaire included a section on drug treatment within the past year (NIJ, 2003). Arrestees were asked about drug treatment in inpatient and outpatient treatment facilities. This information was obtained for each drug and each month for the last 12 months (NIJ, 2003). Although the data obtained from the ADAM sample are now said to be representative of the population of booked arrestees within the catchment areas, problems and limitations with sampling arrestees and other populations have likely led to limitations on the representativeness of the data, a number of which are similar to limitations of the DUF data. These limitations that apply to both DUF and ADAM and their implications will now be discussed.

METHODOLOGICAL ISSUES OF DUF AND ADAM

The DUF and ADAM programs had similar goals. The redesign of the DUF program and the implementation of ADAM were meant to improve the program and increase the quality of the data. Although the program made every effort to meet these goals, the problems associated with interviewing arrestees across the United States posed serious limitations. The major goal of ADAM was the identification of the drug use prevalence and patterns among arrestees. The goal was to produce results that were representative of the arrestees within the catchment area. This may not be the case, however. Even though ADAM used a sophisticated, probability-based sampling method to determine the sites within counties and the persons to be interviewed within sites, the total sample was determined by the cost structure of each site (Yang, 2004). Thus, similar to DUF, sampling in the ADAM program was not uniform for all jurisdictions. Some sites excluded booking facilities with a low volume (e.g., Cleveland). Other sites interviewed arrestees in only a few of their several booking facilities (Birmingham). Additionally, some sites only interviewed felony offenders (e.g., Chicago) and males (e.g., Sacramento) (NIJ, 2003). For both DUF and ADAM, it is likely that the exclusion of booking facilities with certain characteristics influenced the results across sites because different booking facilities might have different arrestees with regard to demographic characteristics and drug use (Yacoubian, 2000).

Availability of Arrestees. Additionally, both DUF and ADAM were only able to sample arrestees who were held long enough in the facility to be interviewed. This is typically true for more serious offenders and indigent offenders who do not have the financial means to make bail. Research suggests that felony offenders use different drugs (crack, cocaine, heroin) than misdemeanor offenders (marijuana, ecstasy). For instance, Webb and Delone (1996) found that felony arrestees were significantly more likely to test positive for cocaine use. There is also a substantial amount of research showing that indigent individuals use different drugs as compared to individuals with greater financial means. A study by Peters, et al. (2002) using the ADAM data suggested that arrestees living under the poverty line were more likely to use opiates and benzodiazepines as compared to arrestees living above the poverty line.

Additionally, crack cocaine is much more widespread among poor black drug users than any other population group. For instance, Hartley, Maddan, and Spohn (2007) found that among black persons facing drug charges, 85.3% used crack cocaine. Among white offenders, only 5.8% faced charges for the possession and/or sale of crack cocaine. White offenders were more likely to use powder cocaine. Other research supports the notion that crack users are disproportionately more often minorities and poor as compared to users of other drugs (Beckett, Nyrop, and Pfingst, 2006).

Although the ADAM weighting procedure attempted to ensure that the different types of offender populations were represented equally in the sample, the weighting procedure was not able to solve the problem associated not having access to certain offender groups because they are released immediately or who don't make it to the facility in the first place because the arresting officer decides to issue a citation instead of an arrest (Yacoubian, 2000). These problems associated with collecting data from arrestees were present for both DUF and ADAM. Thus, both samples should contain information that is representative of more serious and indigent offenders. Myrstol and Langworthy (2005) found that for the ADAM data from Anchorage, the realized male arrestee sample was indeed slightly biased towards felony offenders. This might also be similar for the DUF data.

DETERMINING THE SPECIFIC DRUGS USED WITHIN A CERTAIN JURISDICTION

Another goal of ADAM was to determine the specific drugs used within a certain jurisdiction (NIJ, 2003). Yacoubian (2004) suggests that this goal might not have been met due to the short time period during which interviewers collected data. At most facilities, data collection took place 56 days in one calendar year (14 days per quarter). Similarly, DUF also collected data for each quarter of the calendar year. The difference was that DUF used a predetermined number of interviews (for male offenders it was 225). Once the 225 interviews were completed, data collection stopped. When ADAM was redesigned, the assumption of the data collection by NIJ was that if these 56 days were spread out evenly across the calendar year, all types of drugs used would be detected, and the magnitude of a certain drug's representation could be computed. This assumption is somewhat problematic. Although it is possible that the different types of drugs used in a certain jurisdiction might be detected within the interviewing period, the relatively short period of data collection per year might not have allowed ADAM to accurately determine the magnitude of the drug's representation in that jurisdiction (Yacoubian, 2004). For instance, a certain drug could have started to become more popular during the end of the quarterly data collection period and then leveled out before the new data collection period started. Regardless of whether that is the case or not, it is a problem that applies to DUF as well as to

ADAM, which might have resulted in data that is not substantially different.

Table 3.4 presents the sample size for each quarter by site and year. It appears that the DUF sample for each quarter and each site is very consistent Thus, both studies collected data for a certain period of time and a limited number of arrestees.

NUMBER OF INTERVIEWS FOR EACH QUARTER

Some other patterns emerge from the data. In general, the number of interviews conducted during each quarter in the DUF sample is substantially greater than in the ADAM sample. The exception is Phoenix, where the sample size is somewhat greater for the ADAM data. The ADAM program did not collect data for 2001 in Miami and only collected data for the first quarter of 2001 for Dallas. Additionally, the ADAM sample for the first quarter of the San Antonio sample has only 65 cases, which is less than half as many as for the other three quarters.

The DUF data also has some sites that did not collect data for each quarter and year. Portland did not collect data for the fourth quarter in 1997. Also, the DUF sample for three sites (Dallas, Indianapolis, and Phoenix) has no data for one quarter for 1998 and two sites (Miami and San Jose) are missing data for two quarters of 1998. The DUF reports do not include an explanation why some sites did not collect data for a certain quarter. The sample sizes for DUF, however, were somewhat greater than for ADAM (with the exception of Phoenix) which means that the sample size is not a major problem for DUF. The sample sizes for DUF are probably greater because of the sample target number. Each site attempted to collect 225 interviews per quarter for male arrestees. During ADAM, the sites collected interviews within a certain time period. Additionally, due to the fact that the ADAM questionnaire was much longer than the DUF questionnaire not as many interviews could be completed during ADAM.

Table 3.4 presents the sample sizes for each quarter for DUF and ADAM. These sample sizes were aggregated to represent one year. Finally, the analysis includes two years for DUF and for ADAM. The data for the two years for DUF were aggregated and the data for the two years for ADAM were aggregated, resulting in a fairly large sample for DUF and ADAM. Specifically, the sample sizes for DUF range from 1,272 to 1,959 with the lowest sample size for Miami and the highest sample size for New Orleans. The sample sizes for ADAM ranged from 535 to 2,850. For ADAM Miami had the fewest cases and Phoenix had the most cases. Overall, both samples have a large enough sample size for the analysis (as will be described in the following chapter).

Table 3.4. Sample Size for Each Quarter by Site and

Quarter	DUF		ADAM	
	1997	1998	2000	2001
Dallas				
Q1	247	199	182	178
Q2	240	0	266	0
Q3	245	193	0	0
Q4	248	175	76	0
Year	980	567	524	178
TOTAL	1,547		702	
Denver				
Q1	227	248	125	175
Q2	241	247	150	171
Q3	244	215	166	173
Q4	240	246	177	182
Year	952	956	618	701
TOTAL	1,908		1,319	
Indianapolis				
Q1	224	231	135	191
Q2	247	239	146	180
Q3	241	0	173	175
Q4	225	138	176	186
Year	937	608	630	732
TOTAL	1,545		1,362	
Miami				
Q1	222	219	156	0
Q2	196	199	182	0
Q3	219	0	197	0
Q4	217	0	0	0
Year	854	418	535	0
TOTAL	1,272		535	

Table 3.4. (Continued) Sample Size for Each Quarter by Site and

Quarter	DUF		ADAM	
	1997	1998	2000	2001
New Orleans				
Q1	246	246	147	158
Q2	249	247	156	154
Q3	249	230	150	157
Q4	250	242	151	163
Year	994	965	604	632
TOTAL	1,959		1,236	
Phoenix				
Q1	247	195	251	411
Q2	250	195	331	381
Q3	248	0	339	389
Q4	238	238	345	403
Year	983	628	1,266	1,584
TOTAL	1,611		2,850	
Portland				
Q1	243	215	101	190
Q2	253	212	155	166
Q3	149	178	165	181
Q4	0	151	180	157
Year	645	756	601	694
TOTAL	1,401		1,295	
San Antonio				
Q1	230	228	65	124
Q2	234	227	124	163
Q3	230	228	153	137
Q4	237	226	178	178
Year	931	909	520	602
TOTAL	1,804		1,122	
San Jose				
Q1	215	225	124	183
Q2	221	209	127	166
Q3	214	0	141	186
Q4	235	0	134	187
Year	885	434	526	722
TOTAL	1,319		1,248	

PARTICIPANTS ARE VOLUNTEERS

Most importantly, similar to DUF, the ADAM sample also consists of volunteers because self-selection bias applies to ADAM as well. Even though a probability sample is drawn from the persons arrested, arrestees still had the choice whether to agree to the interview and the urine sample or not. This means that even a probability sample is made up of volunteers, which may differ in their characteristics and drug use from persons who denied participation in the study. For ADAM, refusal rates varied considerably across sites, with a low of 5.9% in Fort Lauderdale to a high of 40.1% in the Charlotte-Metro area. This data was not available for DUF. However, the annual reports from the DUF years 1997 and 1998 estimate the refusal rate to be around 10% for the interviews. It can be expected that the refusal rate for DUF also varied depending on the site. It is possible that the refusal rates of DUF were similar to ADAM because both studies collected data on a sensitive topic in a comparable setting (jail) from booked arrestees. Unfortunately, this question cannot be assessed in depth due to a lack of data for the DUF study. It is likely, however, that self-selection bias might have influenced the data collected in both studies

Also, a number of arrestees did not provide urine specimens. NIJ estimated that approximately 20% of all interviewed arrestees in the DUF program did not provide a urine sample (U.S. Department of Justice, 1998). For the ADAM sample, refusal of urine analysis ranged from a high of 25.3% in Albany/Capital Area, New York to a low of 2.1% in Oklahoma City, Oklahoma. The average refusal rate was 10.8%. Research suggests that arrestees who refused to provide a urine specimen are more likely to be drug users compared to those who agreed to the urine sample (Chen, Stephens, Cochran, and Huff, 1997). This causes an underestimation of drug use among offenders and influences the results and thereby the policy implications drawn from such results. To reiterate, this shortcoming is similar for DUF and ADAM, and for other studies that use bioassays.

LIMITATIONS OF URINE TESTING

Also applicable to both DUF and ADAM are the limitations of urine testing. First, drug testing via urine analysis can only assess drug use for the last 2-4 days, depending on the drug. Second, some drugs are harder to detect than others, which means there is an unknown amount of false negatives. ADAM improved the quality of the urine testing procedure by having all urine samples shipped daily to a central lab. This allowed for an immediate processing of the urine specimen and the testing was completed at the same lab, as a result providing consistent testing. In comparison, during the DUF program, urine samples were shipped to the lab weekly or, at some sites, the urine samples were processed in the lab on site. Riley, Lu, and Taylor (2000) examined the impact of the differential shipment procedures and found that the drug testing results were very consistent. Specifically, the concordance rate was 99%. Thus, it can reasonably be believed that the delay in shipping did not significantly change the drug testing outcomes. For an easier overview, Table 3.5 summarizes the methodology, procedures, and response rates for the DUF and ADAM data.

In sum, there are several limitations that are applicable to both DUF and ADAM that can potentially have a substantial influence on the drug use information obtained from the arrestees. The crucial question is whether these limitations result in similar data in the sense that the drug use information contained in both datasets is not substantially different. Drawing on the two studies conducted by NIJ (1990) and Myrstol and Langworthy (2005) this possibility cannot be ruled out. Based on the limitations faced by both studies and previous research it is hypothesized that the drug use estimates contained in DUF and ADAM are not substantially different despite differences in the sampling method.

The following chapter will describe the data and research strategy used to assess the current research question. Due to the fact that the analysis strategy employed in the current study is rarely used in social sciences, the analytic strategy will be explained by using data from Dallas, Texas, as an example. The results of each step and how they lead to the next step will be discussed. Then, the results for all nine sites will be presented and discussed.

Table 3.5. Comparison of DUF and ADAM

Variable	DUF 1997/1998	1998/Q1,Q2 1999	ADAM 2000/2001
Annual Cost	appr. $1 million		appr. $8.4 million
Sampling Method	Judgment	Judgment	Probability sample
Standardization of Sites	No	Yes	Yes
Number of Sites			
Beginning Number	12	36	39
Final Number	23	36	40
Representative for U.S.	No	NO	No
Selection of Booking Facilities			
Representative for U.S.	No	No	NO
Consistent Catchment Area	No	No	Yes
Facility-level sampling			
Sample Size	Target number of interviews	Target number of interviews	Proportionate to persons arrested
Inclusion/Exclusion Criteria	Priority charge system Twenty percent rule	Priority charge system Twenty percent rule	Random sample from booking lists
Urine			
Analysis across sites	Central Lab and at Sites	Central Lab	Central Lab
Shipment Procedures	Every two weeks	Daily	Daily
Number of Drugs	Between 3 and 10	Between 3 and 10	10
Response Rate			
Interview	90%	90%	56%
Urine Specimen	80%	80%	91%

CHAPTER 4

Data Description and Analytical Strategy

MAJOR RESEARCH QUESTION OF TE CURRENT STUDY

To reiterate, the major research question of the current study is: Are the drug estimates of the Drug Use Forecasting Program (DUF), using a non-probability sample, and drug estimates of the Arrestee Drug Abuse Monitoring Program (ADAM), using a probability sample, substantially different or are they similar enough that they can be said to be equivalent? This research question is based on the criticisms of the sampling procedure of the DUF study by the GAO, that is, it is unknown whether the drug use data contained in the non-probability sample of DUF can be said to be representative of the drug using behaviors among arrestees. This question has not been widely explored. Yet, researchers have used both data sets, DUF and ADAM, to conduct cross-sectional and longitudinal research and make policy recommendations. Assuming that the probability sample of ADAM resulted in a sample and drug use information that can be generalized to the population of arrestees in that specific geographic area (catchment area) the drug use information in the ADAM data will be used as the baseline data and compared to the DUF data.

DATA

The datasets used in this study were obtained from the Interuniversity Consortium for Political and Social Research (ICPSR) and the National Institute of Justice. For DUF, data was available from 1988 until 1998. For ADAM, data was available for the years 2000 and 2001. The DUF data includes data for males and females for the years 1988 - 1997 and for males, females, boys, and girls for the years 1994 till 1997. The ADAM data includes data for males and females for 1998 and 1999 and for males only for 2000 and 2001. The current study will only use data on male arrestees because those data are available for all years for DUF and ADAM and because the ADAM female sample was not a probability sample (NIJ, 1998). Both DUF and ADAM utilized three data sources: (1) arrest records, (2) face-to-face interviews, and (3) urine specimen (NIJ, 1998).

YEARS INCLUDED IN THE ANALYSIS

For the analysis, the ADAM data includes the years 2000 and 2001 and the DUF data includes the years 1997 and 1998. The most appropriate years would be 1998 and 1999 for DUF because those would be the closest to the ADAM years. It is important to use the most consecutive years for the analysis because changes in drug use prevalence and patterns over time could bias the analysis. Research from DUF and ADAM has shown that the popularity of a certain drug varies over time. For instance, Golub and Johnson (2001) demonstrated that marijuana use among arrestees changed substantially between 1991 and 1996. Specifically, in 1991 about 25% of youthful arrestees used marijuana, whereas five years later 57% reported the use of marijuana—a significant increase of 30%. Additionally, Golub and Johnson (1997) found that crack cocaine use declined in Fort Lauderdale from 50% in 1987 to 19% in 1990. This means that within three years crack use dropped by 31%. Using the most consecutive years is thus crucial for the current analysis. Similarly, a study by Yacoubian, et al. (2004) implied that methamphetamine use dropped by more than 50% between 1995 and 1996 for several cities included in the DUF program.

Other drug use surveys also support this result. The NSDUH demonstrates that in 1997 approximately 32.9% of persons 12 years or older had used marijuana in their lifetime and 5.1% had used it within the past month. In 2001, 34.2% of persons 12 years or older had used marijuana in their lifetime and 4.8% had used it within the past month. This demonstrates that the percentage of persons who had "ever used marijuana" increased by 1.3%, but marijuana use in the past month decreased by .03% (NHSDA, 1997, 2001).

These examples show that drug use prevalence is not stable even within a few years. This is of great importance for the current study because the goal is to determine whether the drug use information contained in DUF and ADAM is comparable. Natural fluctuations in the data can have a great impact on the findings. Thus, it is crucial to use the most consecutive years in the current analysis. The year 1999 cannot be included in the analysis, however, because it represents a hybrid year. Part of 1999 was a DUF year in which data was collected from a non-probability sample, but during the latter part of 1999 data was collected with the newly implemented probability sampling method. The year 1999 is divided into quarters 1 and 2 for DUF and quarters 3 and 4 for ADAM because the ADAM report from 1999 states that the new probability sampling procedure was established in 1999. The main problem is that there appears to be no consistent date at which the probability sample was implemented for each site. The reports states: "in the third and fourth quarters of 1999, ADAM sites began implementing new sampling procedures" (NIJ, 1999, p. 7). This statement implies that the realization of the probability sampling procedure did not occur at the beginning of the third quarter at all sites. Rather, the probability sample was established at different times across sites. Also, the 1999 data does not include face sheet data as is provided starting in 2000. The face sheet data shows the basic demographic data for all arrestees selected into the sample. This allows for a comparison of the arrestees who refused or could not be located to those arrestees interviewed. Thus, the lack of face sheet data and the statement in the 1999 report imply that 1999 is not an appropriate year for the purpose of the current study because it is important that the comparison is conducted between the non-probability sample and the probability sample to ensure the validity of the analysis. Thus, the analysis will consist of 1997 and 1998 for DUF and 2000 and 2001 for ADAM.

VARIABLES USED IN THE ANALYSIS

As described in the previous chapter, the main goal of DUF and ADAM was to collect data about drug use prevalence and patterns among arrestees. Thus, the current analysis will focus on the variables measuring drug use. The current study will also examine the demographic profile of DUF and ADAM to assess whether differences in the demographic profile also reflect in differences in drug estimates. Table 4.1 presents and defines the demographic variables use in the study. Table 4.2 shows the codings and descriptions of the drug use variables included in the current analysis.

Table 4.1. Codings & Descriptions for Demographic Variables

Variable Name	Variable Description	Coding
Race	Race of Arrestee	Black = 1; White = 2; Hispanic = 3; Other = 4; Not obtained = 99
Employment	Employment Status of Arrestee	Full Time = 1; Part Time = 2; Unemployed = 3; Other = 4; Not obtained = 99
High school Graduate	Arrestee has Graduated from High school	Yes = 1; No = 0
Offense Category	Arrestees Highest Charge	Violent = 1; Property = 2; Drug = 3; Other = 4
Age	Arrestee Age at the Time of the Offense	>18

Table 4.2. Codings & Descriptions for Drug Use Variables

Variable Name	Variable Description	Coding
MJ	Urine Test Result for Marijuana	Yes = 1; No = 0
COC	Urine Test Result for Cocaine	Yes = 1; No = 0
OP	Urine Test Result for Opiates	Yes = 1; No = 0
MJ72	Self-Reported Marijuana Use Within the Past 72 Hours	Yes = 1; No = 0
COC72	Self-Reported Powder Cocaine Use Within the Past 72 Hours	Yes = 1; No = 0

Table 4.2. (Continued) Codings & Descriptions for Drug Use Variables

Variable Name	Variable Description	Coding
CRK72	Self-Reported Crack Cocaine Use Within the Past 72 Hours	Yes = 1; No = 0
HER72	Self-Reported Heroin Use Within the Past 72 Hours	Yes = 1; No = 0
PCP72	Self-Reported PCP Use Within the Past 72 Hours	Yes = 1; No = 0
AMPH72	Self-Reported Amphetamine Use Within the Past 72 Hours	Yes = 1; No = 0
BARB72	Self-Reported Barbiturate Use Within the Past 72 Hours	Yes = 1; No = 0
EVERCOC	Self-Reported Drug Use – Ever Used Powder Cocaine	Yes = 1; No = 0
EVERCRK	Self-Reported Drug Use – Ever Used Crack Cocaine	Yes = 1; No = 0
EVERHER	Self-Reported Drug Use – Ever Used Heroin	Yes = 1; No = 0
MJALL	Urine Test Result and Self-Reported Drug use for Marijuana (Combines MJ, MJ72, and EVERMJ)	
COCALL	Urine Test Result and Self-Reported Drug use for Cocaine (Combines COC, COC72, CRK72, EVERCOC, and EVERCRK)	
OPALL	Urine Test Result and Self-Reported Drug use for Opiates (Combines OP, HER72, and EVERHER)	

The differences and similarities of the demographic profile and the drug use frequencies will be compared based on the percentage change between the DUF years 1997 and 1998. The percentage difference represents the difference as a percentage of the baseline value (ADAM data). To calculate the percentage difference the difference between the two proportions is calculated and then divided by the baseline value.
The formula for calculating the percentage difference of two proportions is:

%Diff = (p1-p2)/p1

For example, 32.7% of the ADAM arrestees and 43.3% of the DUF arrestees tested positive for marijuana. The calculation would be:

%Diff = (.434 - .327)/.327
%Diff = .327*100
%Diff = 33%

The percentage difference of the demographic variables for the DUF and ADAM data can be considered a straightforward way to get a preliminary overview of the magnitude of differences between DUF and ADAM.

To further assess the question of whether the drug use information contained in the DUF and ADAM samples are substantially different, the current study will conduct equivalence testing, an analysis technique typically used by clinical researchers who are examining whether two drugs or treatments produce outcomes that are not substantially different, meaning that this difference would be of clinical importance. Equivalence testing has been used for a long time by medical researchers and in clinical trials. Since it was first introduced for the use in psychological research by Rogers, et al. (1993) equivalence testing has also become more widespread in other fields as well (Hersen and Gross, 2008). For instance, Epstein, et al. (2001) employed this technique to assess whether web surveys would produce results similar to the traditional paper-and-pencil surveys. Leff, et al. (2005) employed equivalence analysis to compare the quality of services across different health care providers. The following section will now in detail describe equivalence analysis, why it is appropriate for the current study, what the possible outcomes are, and how they will be interpreted. Additionally, since this type of analysis is very rarely used by social scientists, an example will be presented for a better understanding of the statistical technique.

INTRODUCTION TO EQUIVALENCE ANALYSIS

As stated earlier, the main goal of this study is to assess whether the drug use information in DUF and ADAM are substantially different or whether they can be said to be equivalent. Research suggests that equivalence testing is an appropriate method when comparing outcome measures for two different groups or samples (Hauck and Anderson, 1986; Rogers, et al. 1996; Stegner, Bostrom, and Greenfield, 1996; Tryon, 2001).

Equivalence testing has been used extensively in biomedical research where the main goal is to determine whether two drugs or treatments produce equivalent outcomes (Cleophas, Zwinderman, Cleophas and Cleophas, 2009). For this purpose, the researcher compares the outcomes and side effects of the new drug/treatment to the established drug/treatment. Equivalence testing is based on the underlying assumption that two treatments or drugs will always lead to some differences in the outcome. The important question is whether these differences in the outcome are of clinical and/or practical importance (Pocock, 2003). For instance, researchers are introducing a new cancer treatment that has fewer side effects. The new treatment with the fewer side effects is only useful, however, if it is equally effective in treating the illness as compared to the standard treatment. Thus, the goal is to assess the equivalence of these two drugs with regard to the treatment effect or outcome.

Similarly, the current study seeks to determine whether the drug use information in DUF and ADAM are equivalent despite differences in the sampling methods. For the purpose of the current study, the ADAM data will be considered to be the "established treatment" and DUF as the "new treatment." Whereas clinical studies aim to assess the effect a new treatment has as compared to the standard treatment, the current study examines whether the profile of percentage outcomes (the effect) of 14 drug use variables is equivalent for DUF and ADAM. Equivalence does not mean "exactly the same," rather it refers to the "absence of a meaningful difference" (European Medicines Agency, 2000; Rogers, et al. 1993; Allen and Seaman, 2006; Tryon and Lewis,

2009). Stated differently, the question is whether the profile of percentage outcomes of the 14 drug use variables is "comparable" (Hersen and Gross, 2008). Hersen and Gross (2008) suggest that the determination of whether effects are comparable should be made based on "real world outcomes" (p. 216).

Equivalence can be assessed by constructing the confidence interval for proportions, in the current study drug use proportions. Specifically, an equivalence limit or margin (-Δ to +Δ) is chosen defining how different the groups or samples can be before the difference is of practical importance (European Medicines Agency, 2000; Tryon, 2001; Wiens, 2001). After defining the equivalence margin, the two-sided confidence interval is calculated, representing the range of differences between two samples. If the two-sided confidence interval lies within the equivalence limit, then the two groups or samples can be said to be equivalent or "comparable" (Allen and Seaman, 2006; European Medicines Agency, 2000; Hersen and Gross, 2008; Rogers, et al., 1993). The ADAM data will be used as the baseline value and to calculate the equivalence margin because it constitutes a representative probability sample.

Equivalence analysis as described by Rogers, et al. (1993) and Allen and Seaman (2006) consists of two parts: traditional null hypothesis test and equivalence test using the confidence interval approach proposed by Westlake (1981). The researcher can then evaluate the results of these two tests and determine whether the two groups/proportions are substantially different or equivalent. This section will describe how these two tests are carried out and how the outcomes can be interpreted.

TRADITIONAL NULL HYPOTHESIS TESTING

Examining the differences between two groups has traditionally been based on null hypothesis testing. Null hypothesis testing might, however, not be the best strategy when comparing differences between two groups or samples. In the case of null hypothesis statistical testing (NHST) the researcher tests the hypothesis of "no difference," specifically, whether an observed difference between two groups or samples is due to chance (Harris, 1997). Researchers who fail to find significant differences sometimes conclude that the two groups are equivalent. This is not correct, however. A finding of "no difference" does not show that the two groups or samples are equivalent (Gladstein and Makuch, 1984). Another issue is that a statistically significant difference is not necessarily a difference substantial enough to be of practical importance. However, researchers who do find significant differences often conclude that the two groups are indeed different, regardless of how trivial these differences might be or whether they are of practical importance (Rogers, 1996, Tryon, 2001).

Another issue with traditional null hypothesis testing is that it can be overly conservative, increasing the chances of making a Type I Error of rejecting the hypothesis of "no difference". In the current study the z-score will be employed as the test of statistical significance. The associated p-value represents the probability of a Type I Error. A Type I Error means that the null hypothesis will be falsely rejected - that is, the null hypothesis of "no difference" is rejected although it is in fact true and there really is "no difference". For the purpose of the current study, the p-value represents the probability that the researcher concludes that there is a statistically significant difference between the DUF and ADAM data when in fact there is "no difference." The current study uses an α-level of .05, which is a standard level in criminology and other fields. An α-level of .05 means that if the test would be carried out 100 times, 5 of these tests would suggest significant differences, and as a result rejecting the null hypothesis of "no difference" when in fact the null hypothesis is true and there is "no difference" (Moses, 1992).

Accordingly, the hypothesis of "no difference" is rejected if either of the two-tailed tests is significant, that is, the z-value with a 95% confidence interval and an α-level of .05 must be greater than 1.96 or below -1.96 (Agresti and Finlay, 2007). Z-values that lie in the middle of the normal distribution (e.g., in between -1.96 and +1.96) are indicative of the norm, whereas extreme values (values in the tails of the normal distribution) represent the unexpected, something out of the ordinary or "significantly different" (Agresti, 2007). For instance, the p-value for a z-score of 1.97 is .048, which is statistically significant at the .05 level. As the result, the null hypothesis of "no difference" would be rejected. If the z-score is .876, however, the associated p-value would be .38, which is not equal to or lower than .05, and the researcher would not reject the null hypothesis. The equations used for the traditional hypothesis test are shown below:

Calculation of the Traditional Null Hypothesis Test

The traditional z test is computed as:

$z = (p_1 - p_2)/SE$ p = corresponding to the calculated z value

The standard error is calculated as:

$$SE = \left[\frac{(p1)(1-p1)}{n1} + \frac{(p2)(1-p2)}{n2}\right]^{1/2}$$

The 95% Confidence Interval is:

LCL $(p1\text{-}p2) - z_{\alpha/2}(SE)$
LCL $(p1\text{-}p2) + z_{\alpha/2}(SE)$

The current study uses the two-tailed p-value with an α – level of 5% because that is the level used regularly in criminology. Accordingly, the p-value is significant if it is smaller than .025 in the table of the standard normal distribution. In this example, the z-score corresponds to a p-value of .044 in the table of the standard normal distribution. This p-value of .044 constitutes the one-tailed p-value for the null hypothesis. Thus, the two-tailed p-value would be 2*.044 = .088, which is >.05 the predetermined α – level of 5%, indicating that the value is not significant. The null hypothesis of "no difference" cannot be rejected. This result implies that the hypothesis of no difference between DUF and ADAM cannot be rejected. As stated previously, however, this cannot be interpreted as equivalence. To determine whether these two proportions can be said to be equivalent, an equivalence test is conducted simultaneously to the traditional null hypothesis test.

EQUIVALENCE TESTING

In contrast to traditional null hypothesis testing, equivalence testing reverses the null hypotheses. Thus, the null hypothesis is the hypothesis of a "difference," specifically, "the null hypothesis states the difference among group means is greater than some minimal difference representing practical equivalence" (Allen and Seaman, 2006, p.1). For the current study, the null hypothesis will be examined with regard to group proportions, that is, proportions of booked arrestees who used a specific drug. Similar to traditional significance testing, the goal is to reject the null hypothesis. This reversal of the null hypothesis allows the researcher to draw a conclusion of whether there is equivalence between the two groups or proportions (Hauck & Anderson, 1986; Rogers, et al. 1993). Thus, equivalence testing can be seen as a complementary test that allows the researcher to better assess the magnitude of differences (Allen and Seaman, 2006; Hauck and Anderson, 1986; Rogers, 1996). The equations employed for the equivalence test are as follows:

Equivalence Test
Calculation of the Equivalence Test

The equivalence z test is computed as:

$z_1 = (p_1 - p_2) - \delta_1/SE$ p_1 = corresponding to the calculated z-value
$z_2 = (p_1 - p_2) - \delta_2/SE$ p_2 = corresponding to the calculated z-value

The equivalence margin is computed as ±20% of the baseline value (explained below) or stated as an equation: (equals the computation of the percentage difference described above)

$\delta_1 = -20\% * p_1$
$\delta_2 = +20\% * p_1$

The equivalence confidence interval is calculated as:

LCI: $(p_1 - p_2) - z_\alpha(SE)$
UCI: $(p_1 - p_2) + z_\alpha(SE)$
The standard error is calculated as:

$$SE = \left[\frac{(p1)(1 - p1)}{n1} + \frac{(p2)(1 - p2)}{n2}\right]^{1/2}$$

For calculation purposes, only the lower z-value with the higher corresponding p-value needs to be computed (indicating that equivalence does not exist) because it has a greater likelihood of being non-significant than the higher z-value with the lower corresponding p-value. For demonstration purposes, this calculation is shown at the end of this chapter using Dallas as an example.

POSSIBLE OUTCOMES AND INTERPRETATION

Four main outcomes are observed: (1) the drug use information in DUF and ADAM is substantially different (D); (2) the drug use information in DUF and ADAM is equivalent (Eq); (3) the drug use information in DUF and ADAM is different and equivalent (D&Eq); and (4) the results are statistically indeterminate (ND&NEq) (Allen and Seaman, 2006; Rogers, et al., 1993; Tryon and Lewis, 2009).

First, the results are substantially different if the traditional test is statistically significant at the .05 level and if the equivalence test is not statistically significant at the .05 level. Second, the results indicate equivalence if the traditional test is not statistically significant at the .05 level, and the equivalence test is statistically significant at the .05 level. Third, the analysis shows that the two proportions are statistically different but also equivalent. According to some researchers, in this case the difference can be said to be trivial (Allen and Seaman, 2006; Rogers, et al., 1993). The fourth possible outcome is referred to as statistically indeterminate because there is no clear evidence for either statistical difference or equivalence (Allen and Seaman, 2006; Rogers, et al., 1993; Tryon and Lewis, 2009). The results will be said to be statistically indeterminate if the results indicate that the drug use information is neither statistically different nor equivalent. For a better understanding, Figure 4.1 demonstrates how the confidence interval can be used to determine whether there is equivalence between the DUF and ADAM samples.

Figure 4.1. Confidence-Interval Approach for Equivalence Testing (Allen and Seaman, 2006; Rogers, et al., 1993)

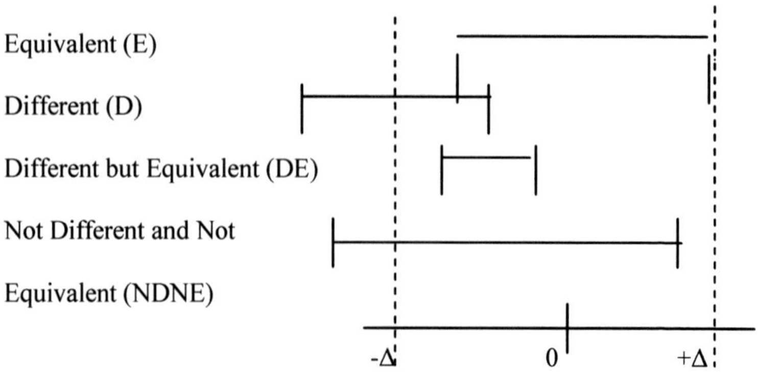

As evident in Figure 4.1, statistical equivalence (E) exists if the confidence interval includes 0 and lies within the equivalence margin (-Δ to +Δ). For instance, the equivalence margin is -24 to +24 and the two-sided confidence interval is -10 and +10. A statistical difference (D) is observed if the confidence interval falls outside the equivalence margin and does not include 0. This would be the case if the equivalence margin is -24 to +24 and the two-sided confidence interval is + 20 and + 46. The results are said to be different but equivalent (DE) if the confidence interval lies within the equivalence margin but does not include 0. Finally, the results are indeterminate (not different and not equivalent) if the confidence interval is not contained within the equivalence margin but does include 0. As shown in Figure 1, this could be the case if the two-sided confidence interval is partially inside and partially outside of the equivalence margin. For instance, the equivalence margin is -24 to +24 and the two-sided confidence interval is - 20 and + 46.

DEFINING THE EQUVALENCE MARGIN

One of the major issues is how to define an appropriate equivalence margin (European Medicines Agency, 2000; Wiens, 2001). Equivalence testing is mostly used in the field of biostatistics and in clinical trials to compare whether two treatments or drugs are equivalent with regard to their effectiveness (Hauck and Anderson, 1986; Wiens, 2002). In the field of bioequivalence studies, the equivalence limit is typically defined as 20% (European Medicines Agency, 2000). Additionally, the European Medicines Agency (2000) states that a 90% confidence interval is an acceptable equivalence interval to evaluate whether the average values of the outcome data are sufficiently close (European Medicines Agency, 2000). Rogers, et al. (1993) suggests that an equivalence interval of 20% is appropriate (p. 557). Equivalence testing has only rarely been used in the social sciences field. For instance, Epstein, et al. (2001) compared the equivalence of internet versus paper-and-pencil assessments. They also used a 90% confidence interval and a 20% equivalence margin. To further assess this issue of an appropriate equivalence interval, drug use research was examined.

To date, there are no set standards in the field of drug use research for the question of what constitutes a substantial difference with regard to changes in the drug use prevalence over time or across different population groups. The term "substantial" is, however, used regularly by researchers to describe differences in the drug use prevalence and patterns. What is substantial differs depending on the drug and the population group being examined. The current study looks at drug use among arrestees, and therefore the drug use literature examining this population group was investigated to determine what is typically considered a substantial change for the drugs included in the current analysis: marijuana, powder cocaine, crack cocaine, heroin, amphetamines, barbiturates, and PCP.

THRESHOLD LEVELS FOR THE MAJOR DRUGS

To reiterate, there exists no clear standard for the question of what constitutes not just a statistically significant change for but also a change that is of practical importance. The current study examines data that includes changes during a time period of five years (1997/1998 compared to 2000/2001). Thus, studies that include trend analysis over several years will be examined as well as reports that assess changes within the last 12 months.

Marijuana. For marijuana, NIJ (1997) reported that for the time period 1996 to 1997 "Marijuana positive rates for juvenile males showed moderate increases (2-8 percentage points) in the majority of sites" (p. 8). Additionally, the DUF report by NIJ (1995) states that there were two sites with "sizable increases" between 1994 and 1995: New Orleans (up 9 points to 16%) and Washington, D.C. (up 8 points to 18%)" (p. 10). This statement implies that an increase of 7% or less is not substantial for a one-year time period. Golub and Johnson (2001) suggest that a percentage change of less than 5% within one year is probably due to random variation. Furthermore, a study by the Australian government examining the impact of legalizing marijuana use on the prevalence rate implies that a percentage of 4% is not substantial (National Drug and Alcohol Research Centre, 1996).

Overall, the examined studies suggest that an increase or decrease in marijuana use of less than 8% is not considered substantial. An 8% increase for marijuana use at a marijuana use rate of 35% corresponds to a percentage difference of 23% (calculated as described above). The annual report published by NIJ (1999) defines a substantial increase to be at least 10% (marijuana positive rates) within two years. The median rate of marijuana positive drug tests in 1999 was 39%. Thus, hypothetically, if there is an increase from 29% to 39% (10% increase), that corresponds to a percentage difference of 34%. Thus, with regard to marijuana there seems to be consistency in the finding that a percentage difference of more than 20% is not necessarily considered substantial. Thus, the current study will use a margin (percentage change) of 20% of the baseline value.

Cocaine. The threshold value for cocaine seems to be similar to that of marijuana. This is not necessarily surprising because cocaine is a popular drug that is used at high rates among arrestees. NIJ (1994) reported that there were a number of substantial increases in cocaine use. In their study, substantial was defined as percentage changes of more than 5% within one year. Golub and Johnson (1997) supported a larger threshold level of 10% within one year by arguing that: "A substantial decline of at least 10 percent in the overall rate of detected cocaine/crack use was observed in Cleveland, Dallas, Detroit, Houston, Los Angeles, New Orleans, Philadelphia, San Diego, San Jose, and Washington, D.C." (p. 11). For a three-year period, however, Golub and Johnson (1997) state that a change of 9% is not substantial. Similar to marijuana, a threshold level of 8% would be a middle ground. Again, an 8% increase or decrease at a median use rate of 30% constitutes a percentage difference of 27%. According to these findings, an equivalence margin of 20% does not seem excessively liberal.

Opiates. The annual DUF report from 1995 only highlighted changes in cities with 10% or more (NIJ, 1995), suggesting that percentage changes of less than 10% were not substantial enough to draw special attention. Furthermore, the ADAM report from 1998 noted: "The most substantial declines for females were recorded in Washington, D.C. and Cleveland (24 percentage points in each) followed by Detroit (17.8), San Jose (17.6), Dallas (17.3) and San Diego (16.6). While it is not possible to know the standard error of these figures, variations of this size suggest substantial changes" (p. 10). Again, a threshold level of 10% for a five-year time period appears to be reasonable. The annual DUF report from 1999, however, defines substantial as an increase or decrease of 5%. A 5% increase at a median use rate of about 10% would be a percentage difference of 50%. Thus, a percentage difference of 50% would be considered substantial. Additionally, the authors state that there was no substantial change between 1998 and 1999. The median drug use positive rate in 1999 was 8%, up 1% from 7% in 1998. This 1% increase results in a percentage difference of 12%. Similar to marijuana and cocaine, an equivalence margin of 20% seems to be reasonable for opiates.

Barbiturates, Amphetamines, and PCP. These three types of drugs are used much more rarely than marijuana, cocaine, or opiates. For instance, amphetamines were used by less than 3% of arrestees across sites. Similarly, barbiturates and PCP were typically used by only 1% of the arrestees. As a result, very small increases or decreases will result in a large percentage difference, and as a result smaller changes are considered substantial. For instance, an increase by 0.6% from 0.2% to 0.8% constitutes a percentage change of 65%.

NIJ (1999) implies that for methamphetamine, a change of 3% can be considered substantial. NIJ (1995) states that a 5% change from 3% to 8% represents a "sharp increase" for methamphetamine. Methamphetamine is, however, used at higher rates than other amphetamines, barbiturates, or PCP. Thus, the threshold for these three drugs would be lower. A percentage difference of 20% (as proposed for marijuana, cocaine, and opiates) would translate into a change of about 0.2% at a use rate of about 1.2%, which appears to be average for the current data. In order to get a better understanding of what a 20% equivalence margin means in absolute numbers, a series of examples will be considered.

SENSITIVITY TEST
UNRINE ANALYSIS TEST OF MARIJUANA

positive tests of marijuana use in Dallas, Texas. The baseline proportion for the urine analysis test for marijuana use in the ADAM data is .327 or 32.7%, with a sample size of 802 arrestees. In actual numbers this means that 262 arrestees tested positive for marijuana. The following three equivalence margins will be considered: 5%, 10%, and 20%.

First, an equivalence margin of 5% would result in an increase or decrease of 1.6% (5%*.327), which is 4 arrestees (1.6%*262)/100. According to the drug use and epidemiological literature, a change of 1.6%is not a substantial change for marijuana use. Second, an equivalence margin of 10% would result in an increase or decrease of 3.3% (10%*.327), which equals 9 arrestees (3.3%*262)/100. A change by 3.3% for marijuana is also not considered substantial. Third, an equivalence margin of 20% corresponds to a change of about 6.5%, or 17 arrestees. According to the literature, a change of about 7% to 8% is considered substantial. Thus, the 20% equivalence margin appears to be a rather conservative measure.

UNRINE ANALYSIS TEST OF OPIATES, DALLAS

Next, these sensitivity tests were repeated for opiates, which have a lower use rate than marijuana and cocaine. Specifically, only 4.2% (34) of the arrestees in Dallas tested positive for opiates in 2000/2001 (ADAM). A 5% equivalence margin is equal to a change of 0.2% or 1 arrestees (exact: .07). A 10% equivalence margin represents a change of .04 or 1 arrestee (exact: 1.4 arrestees). A 20% margin shows a change of 0.8% or 3 arrestees (exact: 2.7 arrestees). Similar to marijuana, a 5% and 10% equivalence margin do not correspond to a change in drug use prevalence that is considered substantial and important according to the literature.

SELF-REPORTED DRUG USE OF PCP WITHIN72 HOURS, DALLAS

PCP is used very rarely. In the current data, only 0.6% of arrestees (or 5 arrestees) reported having used this drug within the past 72 hours. In that case, a 5% equivalence margin means that there would be a change of .03% or .0015 arrestees. A 10% equivalence margin relates to a change of .06% or 0.3 arrestees. Accordingly, a 20% equivalence margin equals a change of .12% or 0.6 arrestees. Again, the 5% and 10% equivalence margin do not constitute a substantial difference. Thus, the 20% margin will be used.

POWER ANALYSIS

Both, the traditional test and equivalence test are algebraically similar to the "independent samples t-test," which tests whether the means or proportions of two groups are statistically different from each other. As a result, the statistical power is also similar (Tryon, 2001). According to Cohen's (1992) "Power Primer" the recommended sample size for an independent samples t-test in order to find a small effect for differences in proportions is N = 392 for an α-level of .05 and N = 584 for an α-level of .01. The sample sizes in the current study are larger than these recommended sample sizes, including Miami which only has data for one ADAM year. Thus, there is sufficient power to conduct the traditional null hypothesis and the equivalence test.

After having discussed why equivalence testing is an appropriate approach for the current study, that there is enough power to conduct the analysis, and how the equivalence margin will be defined, the following part shows the calculation of the equivalence test using the data from Dallas.

ANALYSIS OF DALLAS, TEAXAS DATA

For an easy overview, a summary of the key analysis steps is shown below:

Step 1: Two simultaneous one-sided tests (using the equations provided above)

a) Traditional Hypothesis Test
The goal is to reject the null hypothesis of "no difference" or stated differently: "reject the null hypothesis asserting that the difference between two means (or proportions) is less than or equal to the smaller delta (δ_1)" (Rogers, et al., 1993, p. 554).

Equivalence Test
The goal is to reject the null hypothesis of a "difference" or stated differently: "reject the null hypothesis asserting that the difference is greater than or equal to the larger delta (δ_2)" (Rogers, et al., 1993, p. 554)

Step 2: Evaluation of the Outcome

Four Possibilities:

(a) DUF and ADAM are equivalent (E)
(b) DUF and ADAM are different (D)
(c) DUF and ADAM are different but equivalent (ED)
(d) DUF and ADAM are not different and not equivalent (NDNE)

Analysis
Example: Self-reported marijuana use in Dallas, TX ("Ever Used Marijuana")

Step1: Two simultaneous one-sided tests (using the equations provided above)

a) Traditional Hypothesis Test
p_1 = 68.5% or .685, n = 802 (ADAM)
p_2 = 78.3% or .783, n = 1547 (DUF)

Standard Error:
$SE = [(.685(1-.685))/802 + (.783(1-.783))/1547]^{1/2}$
SE = .019

Traditional 95% confidence interval

LCL: (.685 - .783) – 1.96(.019) = -.136
UCL: (.685 - .783) + 1.96(.019) = -.060

Z-test

z = (.685 - .783)/.019
z = -5.035 p = .000 (one-tailed)

The z-score when using a 95% confidence interval is below 1.96 and thus the p-value is below the .025 level, implying a statistically significant difference for this specific drug variable. Thus, the null hypothesis of "no difference" can be rejected. As described previously, this is not sufficient to conclude that the two proportions are substantially different. In order to draw that conclusion it has to be shown that the two proportions are statistically different and not statistically equivalent. Thus, next the equivalence test is computed.

b) Equivalence Test

Standard Error:
SE = .019 (as calculated previously)

Equivalence margin 20%: (equals the computation of the percentage difference described earlier)

$$\delta_1 = -20\% * .685 = -.1370$$
$$\delta_2 = +20\% * .685 = .1370$$
$$\delta = \pm .1370$$

The equivalence interval would be -.1370 to +.1370 or 13.7%. Next, the equivalence confidence interval will be calculated to see whether it is contained within this equivalence margin.

Example 90% Equivalence Confidence Interval:
LCL: (.685 - .783) - (1.645)(.019) = -.130
UCL: (.685 - .783) - (1.645)(.019) = -.066

The confidence interval does not include 0 but it falls inside the equivalence margin of ±.1370.
Figure 4.2 shows the equivalence margin and the confidence interval.

Figure 4.2. Example Equivalence Test

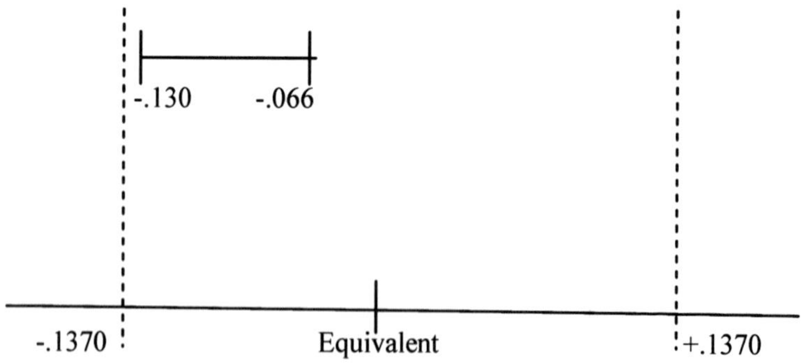

To further assess whether substantial differences exists, the equivalence z will be computed. If both tests have a p-value with a α – level of .05 or less the null hypothesis of a "difference" can be rejected and the two proportions can be said to be equivalent. As described above, only the lower z-score with the higher p-value needs to be computed because it is more likely to be not statistically significant than the higher z-score with the lower p-value.

Example Equivalence z:
$z_1 = ((.685 - .783) - (-.685))/.019 = 2.004 \quad p = .000$

This finding indicates that the z-test with the higher p-value is significant for the one-tailed test at the .05 α – level because the z-score when using the 90% confidence interval is greater than 1.645. Accordingly, the null hypothesis of a "difference" can be rejected and equivalence can be concluded. This example implies that the DUF and ADAM data are statistically different but statistically equivalent. The following section will now present the results of the current study.

Decision Rules

At this point it is important to lay out the criteria for the interpretation of the results and the hypotheses. As described at the beginning of this Chapter, there are four possible outcomes:

(1) DUF and ADAM can be said to be equivalent (Eq)
(2) DUF and ADAM are different (D)
(3) DUF and ADAM are different and equivalent (D&Eq)
(4) DUF and ADAM are not different and not equivalent (indeterminate) (ND&NEq)

To reiterate, the main question is whether the drug use estimates of DUF and ADAM are substantially different. Two of the four possible outcomes are straightforward with regard to their interpretation. If the variable is statistically different and not statistically equivalent (D), then the variable is classified as substantially different. If the variable is statistically equivalent and not statistically (Eq) different, it can be classified as not being substantially different. The other two outcomes are somewhat more ambiguous. If the variable is statistically different and statistically equivalent (D&Eq), it can be classified as not substantially different because even though the traditional null hypothesis test showed a statistically significant difference this difference was marginal in the sense that the confidence interval of the proportion still falls within the equivalence margin. Finally, if the variable is not statistically different and not statistically equivalent (ND&NEq), it cannot be classified as either substantially different or not substantially different. These variables are statistically indeterminate.

It is to be expected that each site will have several of these outcomes and thus there is the issue of how these multiple findings are to be interpreted. For example, for Dallas 27% (3) of the variables fell into the category *equivalent*, 27% (3) were statistically *different but statistically equivalent* (D&Eq), and 45% (5) of the variables were *not statistically different and not statistically equivalent* (ND&NEq) and cannot be judged because they were statistically indeterminate. The following decision criteria are proposed to judge these results.

First, for the purpose of the current study the findings of *Equivalent* (Eq) and *Different but Equivalent* (D&Eq) will be combined and interpreted as not substantially different. Researchers suggest that a finding of *Different but Equivalent* (D&Eq) means that although there is a significant difference using the traditional null hypothesis test these differences might be trivial and not of practical importance. Specifically, Rogers, et al. (1993) states that the "difference was larger than the standard null value (usually zero) but smaller than a difference that would make the groups nonequivalent" (p.561). This interpretation is also supported by Allen and Seaman (2006) who state that "there is a difference, but it is trivial" (p. 78). They suggest that this finding might occur because the study was "overpowered." This term refers to the possibility that with a large sample size it is possible to detect a significant difference when in fact the difference is "scientifically insignificant" (Frank and Althoen, 1994). This type of error (Type I Error) is also one of the main criticisms on traditional null hypothesis testing, namely that with large sample sizes researchers will often find a statistically significant difference (Batanero and Diaz, 2006). Finding a statistically significant difference does not necessarily mean that the difference is of practical importance (Batanero and Diaz, 2006; Levin, 1998; Vacha-Haase, 2001). Conducting the traditional null hypothesis test and the equivalence test simultaneously provides a better understanding of the magnitude of the actual differences. In the current study, the percentage outcomes for certain drugs may be statistically different but at the same time statistically equivalent in the sense that the 90% equivalence confidence interval falls with the equivalence margin. Thus, these outcomes will be interpreted as not substantially different or similar.

Second, sites will be classified as "unable to be assessed" if more than one third of the variables fall into the category "*Not Statistically*

Different and Not Statistically Equivalent." In actual numbers, it means that sites will be said to be unable to be assessed if more than three variables are statistically indeterminate. Rogers, et al. (1993) suggest that such findings might be due to excessive "within group variation" and as a result the equivalence analysis is inconclusive. This is believed to be a conservative approach.

Third, sites will be classified as substantially different if 20% or more of the drug use values show substantial differences. In actual numbers, that means that sites will be determined to be substantially different if three variables or more drug use values are classified as "*Different.*" As described in Chapter 2, drug use prevalence and patterns are not constant over time and a certain amount of changes can be contributed to these naturally occurring differences as long as the differences reflected in the current study are consistent with changes evident from other drug use studies. For example, urine test results for cocaine in Denver are substantially different. The DUF and ADAM data show that cocaine use declined consistently from 40.2% in 1997 to 32.8% in 2002 (see Table B.2., Appendix B). This decrease in cocaine use is supported by the data of the major national drug use surveys.

Similarly, data from the DAWN study shows that emergency room visits due to heroin use have increased by 33% between 1995 and 2001. This is a substantial increase and it can be expected that such increase will also reflect in the DUF/ADAM data. This study includes three measures of heroin use (MJ, MJ72, and EVERMJ), two of which refer to recent use (MJ and MJ72). These two variables measuring recent heroin use would likely be affected by the actual increase of heroin use as shown by the DAWN data.

Data from the Treatment Episode Data Set (TEDS) also shows a decrease in emergency room admissions for cocaine use for Texas. The NSDUH also demonstrates a decline in cocaine use in Texas, albeit a smaller decline (SAMHSA, 2001, Appendix A). Additionally, the MTF suggests that cocaine use among school children declined from about 5.9% in 1997 to 4.8% in 2001. Considering the consistency of these findings with regard to a decline in cocaine use, the substantial difference found for the urine test result for cocaine in Denver might be attributable to an actual change in drug using behaviors rather than the change in the sampling method. Thus, it appears reasonable to expect some variables to be substantially different. Limiting the number of variables to two could be seen as a conservative approach considering the fact that the data cover a five-year time period and there are two measures of recent drug use for each drug - that is, the urine analysis result for marijuana, cocaine, and opiates, and drug use within the past 72 hours for marijuana, cocaine, and heroin.

In summary, the following decision criteria are proposed:

1) The findings of *Equivalent* (Eq) and *Different but Equivalent* (D&Eq) will be interpreted as not substantially different or similar.
2) Sites will be classified as "unable to be assessed" if more than one third of the drug use values fall into the category "*Not Statistically Different and Not Statistically Equivalent*" (ND&NEq).
3) Third, sites will be classified as substantially different if 20% or more of the drug use values show substantial differences (D).

CHAPTER 5

Results of the Equivalence Analysis: Assessing the Impact of Sampling Procedures

This Chapter presents the results of the equivalence analysis. This section will begin by examining the demographic profile for DUF (1997/1998) and ADAM (2000/2001) to get a general overview of the similarities and differences in the two datasets. After the assessment of the demographic profile, the current analysis will take a look at the drug use frequencies for DUF (1997/1998) and ADAM (2000/2001). The purpose of examining the drug use frequencies is twofold: (1) to provide an overview of the prevalence of each drug among arrestees; and, (2) to decide which drug use variables will be excluded from the equivalence analysis. Finally, the results of the equivalence analysis will be presented to assess whether the drug use estimates of DUF and ADAM are substantially different. The results of the equivalence analysis will be presented in two parts: (1) overall results for the equivalence tests; and, (2) site specific results. The decision criteria described in Chapter Four will be used to interpret the findings.

DESCRIPTIVE STATISTICS

DEMGRAPHIC PROFILE

Tables A.1. to A.9. in Appendix A display the demographic profile of the sample for the nine sites included in the current study. The data shows that the demographic profile of DUF (1997/1998) and ADAM (2000/2001) has similarities and differences. The results are presented by demographic category.

Race. The racial distribution is described for each racial category, including Black, White, Hispanic, and Other. The percentage of Black arrestees appeared to be similar for Indianapolis, New Orleans, Phoenix, San Antonio, and San Jose. These sites had a difference of less than 5% between DUF and ADAM. Specifically, in Indianapolis, 57.2% of arrestees were Black in the DUF sample, and 56.5% were Black in the ADAM sample. In New Orleans, 87% of arrestees were Black in the DUF sample, and 86.6% were Black in the ADAM sample. In Phoenix, 13.7% of arrestees were Black in DUF and 11.9% in ADAM. In San Antonio, 11.3% of arrestees were Black in the DUF sample, and 12.7% were Black in the ADAM sample. Finally, in San Jose, 11.1% were Black in the DUF sample, and 11.9% were Black in the ADAM sample.

Four sites had a difference of 6% or more. These sites are Miami, Portland, Dallas, and Denver. The largest difference was apparent in Dallas, where 58.4% of arrestees in the DUF sample were Black, but only 49.1% of arrestees in the ADAM sample were Black. Additionally, the difference in Miami was 6.7%, in Portland 5.4%, and in Denver 8.2%. There was no consistent pattern with regard to the direction of the difference; that is, in Portland, Dallas, and Denver, the DUF sample had a greater number of Black arrestees as compared to the ADAM sample, whereas in Miami the DUF sample had fewer Black arrestees than the ADAM sample.

The findings for White offenders demonstrate that Portland, Indianapolis, New Orleans, Denver, San Antonio, and San Jose have differences of less than 5% between DUF and ADAM. In Portland, the DUF sample included 61.7% White arrestees; the ADAM sample included 64.6% White arrestees. Indianapolis had 37.9% White arrestees in the DUF sample, and the ADAM sample had 42.4%. The DUF sample in New Orleans consisted of 11% White arrestees, whereas the ADAM sample consisted of 12.9%. The Denver site had 28.6% White arrestees in the DUF sample and 27.6% in the ADAM sample. San Antonio included 32.9% White arrestees in the DUF sample and 35.3% in the ADAM sample. Finally, the DUF sample in San Jose consisted of 32.1% White arrestees and the ADAM sample of 30.4%.

Phoenix, Miami, and Dallas showed differences that were larger than 5%. The largest discrepancies were found in Miami, where about 15.7% of arrestees were White in the DUF sample, but 43.2% of arrestees were white in the ADAM sample. The remaining two sites demonstrated a difference of 6.6% (Phoenix) and 6.5% (Dallas). Again, no consistent pattern for the direction of the discrepancies emerges. Phoenix and Miami had fewer White arrestees in the DUF sample and more in the ADAM sample. Contrary, in Dallas the DUF sample included a greater number of White arrestees than the ADAM sample.

The assessment of differences for Hispanic offenders is more complicated because the percentage of Hispanics in the sample varies greatly. Three sites had less than 7% Hispanic offenders in the sample overall. These sites were Portland, Indianapolis, and New Orleans. With regard to the differences between DUF and ADAM, the data shows that in Portland the DUF sample had 5.6% Hispanic arrestees, whereas the ADAM sample had 7.3% Hispanic offenders. Indianapolis had 4.1% Hispanic arrestees in the DUF sample but only 0.8% in the ADAM sample. In New Orleans, the DUF sample consisted of 0.8% Hispanic offenders and the ADAM sample of 0.2%.

The vast majority of sites show substantial differences. The greatest differences were found in Miami, where the DUF sample included 37.7% Hispanic arrestees, but the ADAM sample only had 3.9% Hispanic offenders. Also, in Denver 33.8% of the DUF sample were Hispanic arrestees, but 41.5% of the ADAM sample were Hispanic offenders. In Dallas, 9.8% of arrestees were Hispanic in the DUF sample and 14.8% in the ADAM sample. In Phoenix, 32.3% of arrestees in the DUF sample were Hispanic, and 25.6% of arrestees in the ADAM sample were Hispanic. In San Antonio and San Jose, the percentage of Hispanics in the sample appeared to be similar, with differences of less than 3%. Of the three racial categories, Hispanic arrestees showed the greatest discrepancies between DUF and ADAM. The important question is whether these differences in the racial distribution also lead to differences in the drug use estimates between DUF and ADAM.

Employment. For the percentage of arrestees employed full time, only one site had a difference of more than 5%. That site was Miami, with a difference of 6%. Conversely, the category part-time employment was substantially different for all sites with the exception of Phoenix, where the difference was only 2%. The remaining sites demonstrated differences of more than 6%. Similarly, the percentages of arrestees who stated they were unemployed were also substantially different at all sites. The DUF sample had much fewer arrestees who were unemployed at the time as compared to the ADAM sample.

Specifically, at most sites more than twice as many arrestees in the ADAM sample fell into the unemployed category as compared to the DUF sample.

Education. The frequency distributions also suggest substantial differences with regard to the high school graduation rate of arrestees between the DUF and ADAM samples. The greatest discrepancies were found in San Jose and Miami. In San Jose, the difference was 27.8% because only 49.8% of arrestees in the DUF sample graduated from high school but 77.6% of arrestees in the ADAM sample graduated from high school. Similarly, the difference in Miami was 26.8%; 40.3% of arrestees in the DUF sample graduated from high school as compared to 67.1% in the ADAM sample. The remaining sites had differences of 10% or more. There is a consistent pattern for this category—the ADAM sample has a higher percentage of high school graduates for all sites. Thus, there appears to be a systematic bias in the data for high school graduation rates.

Charge distribution. The charge distribution for DUF and ADAM were expected to be different because of the differences in the arrestee selection procedure. As described in Chapter Four, DUF used a priority charge system, whereas ADAM used the UCR in their determination of the sample. Despite these differences in the arrestee selection procedure, four sites appeared to have a similar charge distribution. Specifically, in Denver, Phoenix, Indianapolis, and Portland, the charge distribution for DUF arrestees was within 5% for all three categories (violent, property, and drug offenses) for that of ADAM. The other five sites demonstrated substantial differences, with discrepancies of 10% and more for each charge category (violent, property, and drug offenses). Again, these differences were to be expected because DUF and ADAM used different methods to determine how many arrestees should be included for each charge category.

Age. Contrary to the charge distribution, the average age of the arrestees in the DUF and ADAM samples was almost exactly the same for all sites. Specifically, for eight sites the average age was the same for DUF and ADAM. The sole exception is San Jose, where the average age of the arrestees in the ADAM sample was 32 as compared to 31 for the DUF sample.

In sum, from the frequency distributions there appear to be a number of similarities between DUF and ADAM but also differences for the demographic profile of the two samples. This is not an unexpected finding because ADAM employed a probability sample and a weighting procedure to ensure that the arrestee sample was representative of arrestees booked in the catchment area, whereas DUF did not use these procedures. The main question is whether these differences influence the drug use estimates produced by DUF and ADAM in a way that makes them substantially different.

DRUG USE FREQUENCIES

Table 5.1. shows the prevalence rates for drug use by presenting the lowest, highest, and average prevalence rates for all sites combined. This data was compiled from Tables B.1. – B.9. in Appendix B, where the prevalence rates of drug use are displayed for each site for each year for DUF (1997 and 1998) and ADAM (2000 and 2001). The drug use data for both DUF and ADAM appears to be similar with regard to the prevalence of the drugs examined.

The most-often used drug is marijuana (MJ). The high prevalence of marijuana shows in both the urine test results and the self-reported data for the DUF and ADAM data. Second, cocaine (COC) has the next highest prevalence rates for the urine test results and the self-report data for DUF and ADAM. The self-report data further shows that the average prevalence rate for crack cocaine (CRK) used within the past 72 hours was higher than for powder cocaine. Contrary, for lifetime use (ever used drug) powder cocaine had higher average prevalence rates than crack cocaine. Again, these patterns are consistent for both DUF and ADAM. Third, in both datasets, opiates (OP), including heroin (HER), had much lower prevalence rates than marijuana and cocaine, but were more popular than PCP (PCP), amphetamines (AMPH), and barbiturates (BARB). Fourth, PCP, amphetamines, and barbiturates were used very rarely by arrestees in the DUF and ADAM samples. These overall findings will be examined in more detail now.

First, the prevalence rates shown in Table 5.1. indicate that in the ADAM sample between 60.2% and 84.1% of arrestees stated that they have used marijuana at some point in their life (EVERMJ). On average, 74.2% of arrestees have previously used marijuana (EVERMJ). Similarly, the DUF data shows that between 63.3% and 87.4% of arrestees have used marijuana (EVERMJ). On average, 76.0% of arrestees admitted to having used marijuana (EVERMJ).

With regard to recent use, the ADAM data implies that between 22.1% and 35.0% of arrestees reported having used marijuana within the past 72 hours (MJ72), with the average being 27.0% (MJ72). The DUF data suggests that between 18.9% and 34% have used marijuana in the past 72 hours (MJ72), with the average being 26.9%. Additionally, the ADAM data demonstrates that on average 38.4% of arrestees tested positive for marijuana (MJ) use. Similarly, in the DUF sample, 36.9% of arrestees tested positive for marijuana (MJ).

Second, cocaine is also used regularly by arrestees. Specifically, the ADAM data demonstrates that between 3.0% and 9.7% of arrestees used powder cocaine within the past 72 hours (COC72) and between 3.3% and 11.2% used crack cocaine within the past 72 hours (CRK72). On average, 5.8% of arrestees had used powder cocaine in the past 72 hours (COC72) and 9.8% had used crack cocaine in the past 72 hours (CRK72). In comparison, the DUF data shows that between 3.2% and 11.3% of arrestees used powder cocaine within the past 72 hours (COC72) and between 2.8% and 17.2% used crack cocaine within the past 72 hours (CRK72). For the DUF data, the average use rates within the past 72 hours were 7.1% for powder cocaine (COC72) and 11.8%% for crack cocaine (CRK72).

The ADAM data also shows that between 27.9% and 50.8% of arrestees have used powder cocaine (EVERCOC) and between 14.8% and 41.1% have used crack cocaine (EVERCRK) at some point in the life. For the DUF data, the lifetime prevalence rates range from 32.3% to 54.1% for powder cocaine (EVERCOC) and 12.2% and 44.5% for crack cocaine (EVERCRK).

Additionally, between 12.0% and 44.9% of arrestees in the ADAM tested positive for cocaine (COC), with the average being 28.9%. For the DUF data, the percentage of arrestees who tested positive for cocaine (COC) ranged from 11.8% to 53.9%, with an average of 33.9%. To reiterate, the urine analysis cannot distinguish between crack and powder. Thus, only overall cocaine use is reported.

Third, opiate use (including heroin use) is rarer among arrestees as compared to marijuana and cocaine. On average, 7.1% of arrestees tested positive for opiates (OP) in the ADAM data and 6.8% in the DUF data. The urine positive tests ranged from 3.5% to 15.1% for ADAM and 2.2% to 14.8% for DUF.

With regard to lifetime heroin use (EVERHER), between 6.6% and 24.2% of arrestees in the ADAM data and between 7.6% and 31% of arrestees in the DUF data reported having used heroin. On average,

14.2% of ADAM arrestees 15.8% of DUF arrestees had used heroin at some point in their life (EVERHER).

The data for recent heroin use (HER72) demonstrates that, on average, 4.4% of arrestees in the ADAM data and 4.5% of arrestees in DUF used heroin within the past 72 hours. Heroin use within the past 72 hours (HER72) ranged from 1.3% to 11.2% for ADAM and 0.3% to 11.0% for DUF.

The descriptive analyses also show that the variables PCP72, AMPH72, and BARB72 have very low prevalence rates. Specifically, the prevalence rates for recent PCP use (PCP72) varied between 0.0% and 1.6% for ADAM and 0.1% to 0.8% for DUF. These percentages are equivalent to a range of 0 to 20 arrestees. On average, PCP72 was used by .04% of arrestees in the ADAM data and 0.3% of arrestees in the DUF data.

The prevalence rate for recent use of amphetamines (AMPH72) ranged from 0.2% to 1.4% for ADAM and 0.0% to 3.4% for DUF. On average, 0.7% of arrestees used amphetamines within the past 72 hours (AMPH72) in the ADAM data and 1.6% in the DUF data.

Similarly, the prevalence rates for barbiturates (BARB72) were also very low, varying from 0.0% to 0.5% for ADAM and 0.4% to 2.0% for DUF. On average, 0.2% of arrestees used barbiturates in the ADAM data and 0.8% in the DUF data. These percentages equal a total number of arrestees between 0 and 16. Due to these very low sample sizes, the variables PCP72, AMPH72, and BARB72 were excluded from the equivalence analysis.

So far, the descriptive analysis has demonstrated that there are differences in the demographic characteristics of the DUF and ADAM samples. The descriptive statistics, however, also show that the overall prevalence of the different drugs included in the analysis appears to be similar. These findings support the findings of the NIJ (1993) study which found that even though the demographic characteristics and charge distribution of the DUF and ADAM samples were substantially different, the drug use estimates were similar. The following section will present the results for the equivalence analysis to determine whether these findings also hold up when a more rigorous statistical test is applied.

Table 5.1. Drug Use Frequencies for Total Sample – Lowest, Highest, and Average Prevalence Rates

Variable	ADAM 2000/2001				DUF 1997/1998			
	Lowest	Highest	Average	N Average	Lowest	Highest	Average	N Average
Urine Test								
MJ	32.7	48.7	38.4	1,308	27.5	44.3	36.9	1,600
COC	12.0	44.9	28.9	1,308	11.8	53.9	33.9	1,600
OP	3.5	15.1	7.1	1,308	2.2	14.8	6.8	1,600
Self-Report Drug Use Within 72 hours								
MJ72	22.1	35.0	27.0	1,308	18.9	34.0	26.9	1,600
COC72	3.0	9.7	5.8	1,308	3.2	11.3	7.1	1,600
CRK72	3.3	12.6	9.8	1,308	2.8	17.3	11.8	1,600
HER72	1.3	11.2	4.4	1,308	0.3	11.0	4.5	1,600
PCP72	0.0	1.6	0.4	1,308	0.1	0.8	0.3	1,600
AMPH72	0.2	1.4	0.7	1,308	0.0	3.4	1.6	1,600
BARB72	0.0	0.5	0.2	1,308	0.4	2.0	0.8	1,600
Ever Used Drug								
EVERMJ	60.2	84.1	74.2	1,308	63.3	87.4	76.0	1,600
EVERCOC	27.9	50.8	38.9	1,308	32.3	54.1	40.3	1,600
EVERCRK	14.8	41.1	35.2	1,308	12.2	44.5	30.4	1,600
EVERHER	6.6	24.2	14.2	1,308	7.6	31.0	15.8	1,600

OVERALL RESULTS FOR THE EQUIVALENCE ANALYSIS

To reiterate, the research question is whether the drug use estimates for selected drugs are substantially different between DUF and ADAM. The main hypothesis is that the drug use estimates contained in DUF and ADAM are not substantially different despite differences in the sampling method. In order to assess the hypothesis, the DUF and ADAM data for the nine sites was compared to determine how many drugs and sites, if any, appear to be substantially different and how many, if any, appear to be similar. First, the drug use estimates for the 11 variables are similar and different for DUF and ADAM for all sites combined will be discussed. Tables 5.2. through 5.4. present this information.

Table 5.2. shows how many variables are substantially different, how many are not substantially different or similar, and how many are statistically indeterminate for all sites combined. Each site had a total of 11 drug use estimates – one for each drug. The sites with the greatest number of substantially different drug use values are Phoenix with 36% (4 values) and Portland with 27% (3 values). In Denver and New Orleans, 18% (2) of the drug use values were classified as substantially different. For the three sites Indianapolis, Miami, and San Jose, 9% (1) of the drug use values were substantially different. There are two sites where none of the drug use values were substantially different. These sites are Dallas and San Antonio. Altogether, out of a total of 99 drug use values (11 variables * 9 sites), 14 values were found to be substantially different.

As explained above, the categories "*Equivalent*" and "*Different but Equivalent*" will be counted as not substantially different or similar. Thus, these two categories will be evaluated together. The highest number of values that were classified as not substantially different was found in New Orleans with 81% (9 values). Specifically, four drug use values fell into the category "Equivalent" and five values were classified as "Different but Equivalent." Several sites had 54% (8) of the drug use values that fell into either the "Equivalent" or "Different but Equivalent" category. These sites were Denver, Portland, San Antonio, and San Jose. In Denver, four drug use values were "Equivalent" and four values were "Different but Equivalent." In Portland and San Jose, three drug use estimates showed "Equivalence" and five values were "Different but Equivalent." Finally, in San

Antonio, six drug use values were classified as "Equivalent" and two values were classified as "Different but Equivalent."

In Indianapolis and Miami, 45% (7) of the values showed no substantial differences. For Indianapolis, four values were "Equivalent" and three values were "Different but Equivalent." In Miami, five values proved to be "Equivalent" and two fell into the category "Different but Equivalent."

Additionally, for two sites, a total of 36% (6) of the drug use values fell into either of these two categories. In Phoenix and Dallas, three drug use values were classified as "Equivalent" and three values fell into the category "Different but Equivalent." No site had less than 36% (6 values) classified as either "Equivalent" or "Different but Equivalent."

Table 5.2. Equivalence Test Outcomes by Site

	Number of Drug Use Variables							
Site	Different		Equivalent		Different but Equivalent		Not Different and Not Equivalent	
	%	N	%	N	%	N	%	N
Dallas	0%	0	37%	3	27%	3	45%	5
Denver	18%	2	36%	4	36%	4	9%	1
Indianapolis	9%	1	36%	4	27%	3	27%	3
Miami	9%	1	45%	5	18%	2	27%	3
New Orleans	18%	2	36%	4	45%	5	0%	0
Phoenix	36%	4	27%	3	27%	3	9%	1
Portland	27%	3	27%	3	45%	5	0%	0
San Antonio	0%	0	55%	6	18%	2	27%	3
San Jose	9%	1	27%	3	45%	5	18%	2
Total Values	14		35		32		18	

Finally, there were also a number of sites with values that fell into the category "Not Different and Not Equivalent." These values are statistically indeterminate and cannot be judged as either substantially different or not. The greatest number of statistically indeterminate values had Dallas, where 45% (5) of the values fell into that category. Three sites had 27% (3) values that cannot be judged: Indianapolis, Miami, and San Antonio. In San Jose, 18% (2) values were indeterminate and in Denver and Phoenix, 9% (1) value was indeterminate. At two sides, none of the values was statistically indeterminate, that is, all drug use values were categorized as either substantially different, equivalent, or different but equivalent. These sites were New Orleans and Portland.

The results for each site as displayed in Table 5.2. gives a general overview of how many drug use values fall into each outcome category. This data does not provide information, however, of whether these outcome categories consist of the same variables across sites. For instance, Dallas, Phoenix, Portland, and San Jose have three drug use values that were categorized as "Equivalent." The question is whether those three values are the same for the four sites or whether each site has different patterns with regard to the question how the drug use values are distributed over the outcome categories. Thus, the next step in the analysis was to examine the distribution of these drug use variables across the outcome categories to get a better understanding of possible patterns in the data. The results are presented in Tables 5.3. and 5.4.

Table 5.3. shows the classification for each drug use value and site. The drug use variables are ordered in their original way, that is, the first three variables are the urine test results, followed by the variables for self-reported drug use within the past 72 hours, and finally, the self-reported lifetime drug use. Table 5.4. shows the same data but summarized for each outcome category and drug use variable. Also, the drug use variables are ordered differently. The first three drugs are the variables for marijuana, followed by the variables for cocaine and finally opiates. This order makes patterns more obvious.

The results show that there are indeed certain patterns with regard to the distribution of drug use values across the outcome categories. First, the marijuana variables fall almost exclusively into the two categories "Equivalent and Not Different" (Eq) and "Different but Equivalent" (D&Eq). Only one value (4%) for all marijuana variables shows substantial differences (D). This value is "Marijuana Used within the Past 72 Hours" in Indianapolis. None of the values for marijuana were classified as indeterminate. Thus, the drug use values

Table 5.3. Classification of Drug Use Values Across the Outcome Categories for Each Site

Site	Drug Use Variables										
	MJ	COC	OP	MJ72	COC72	CRK72	HER72	EVERMJ	EVERCOC	EVERCRK	EVERHER
Dallas	D&Eq	ND&NEq	Eq	D&Eq	ND&NEq	ND&NEq	ND&NEq	D&Eq	Eq	Eq	ND&NEq
Denver	Eq	D	Eq	Eq	ND&NEq	D	D&Eq	D&Eq	D&Eq	D&Eq	Eq
Indianapolis	D&Eq	D&Eq	D&Eq	D	ND&NEq	ND&NEq	ND&NEq	Eq	Eq	Eq	D&Eq
Miami	Eq	D	D&Eq	Eq	ND&NEq	ND&NEq	D&Eq	Eq	Eq	ND&NEq	Eq
New Orleans	D&Eq	D&Eq	D&Eq	Eq	D&Eq	D&Eq	Eq	Eq	D	D	Eq
Phoenix	D&Eq	D	D	D&Eq	D	ND&NEq	D&Eq	Eq	Eq	Eq	D
Portland	Eq	D&Eq	D	Eq	D&Eq	D&Eq	D&Eq	D&Eq	D	Eq	D
San Antonio	Eq	Eq	ND&NEq	Eq	ND&NEq	Eq	ND&NEq	D&Eq	Eq	D&Eq	Eq
San Jose	D&Eq	Eq	D	D&Eq	ND&NEq	Eq	ND&NEq	D&Eq	D&Eq	D&Eq	Eq

Table 5.4. Classification of Drug Use for Each Outcome Category

Site	Drug Use Variables											
	MJ	MJ72	EVERMJ	COC	COC72	EVERCOC	CRK72	EVERCRK	OP	HER72	EVERHER	Total
D	0	1	0	3	1	2	1	1	3	0	2	14
Eq	4	5	4	3	0	5	2	4	2	1	5	35
D&Eq	5	3	5	2	2	2	2	3	3	4	1	32
ND&NEq	0	0	0	1	6	0	4	1	1	4	1	18

for marijuana can be said to show no substantial differences for any site and overall for all sites combined.

Figure 5.1. summarizes the distribution of the marijuana values across the outcome categories. For the purpose of simplification, the outcome categories were recoded to better reflect their meaning. The four original outcome categories were collapsed into three categories. Specifically, the category "Statistically Different and Not Statistically Equivalent" was renamed into "Substantially Different." The outcome categories "Statistically Equivalent and Not Statistically Different" and "Statistically Different but Statistically Equivalent" were collapsed into "Not Substantially Different." Finally, the category "Not Statistically Different and Not Statistically Equivalent" was renamed "Indeterminate." Again, Figure 5.1. very clearly demonstrates that all but one variable show no substantial differences.

Figure 5.1. Distribution of Marijuana Values Across the Outcome Categories

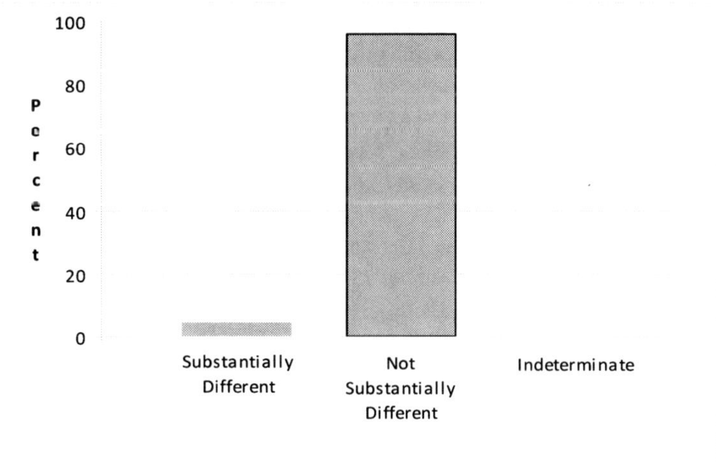

Second, for the drug use variables for cocaine use, the data displayed in Tables 5.3. and 5.4. implies that the results are not as clear cut as for marijuana. The urine test results for cocaine (COC) show that three values were substantially different (Denver, Miami, and Phoenix), five values were not substantially different (Indianapolis, New Orleans,

Portland, San Antonio, and San Jose), and one value could not be statistically determined (Dallas).

For self-reported powder cocaine use within the past 72 hours (COC72), the analysis demonstrates that six values could not be statistically determined (Dallas, Denver, Indianapolis, Miami, San Antonio, and San Jose). Of the remaining three values, one value was substantially different (Phoenix) and two were not substantially different (New Orleans and Portland).

The pattern for self-reported crack cocaine use within the past 72 (CRK72) hours demonstrates that one value was substantially different (Denver), four values were not substantially different (New Orleans, Portland, San Antonio, and San Jose). Finally, four values were classified as indeterminate (Dallas, Indianapolis, Miami, and Phoenix).

The outcomes for self-reported powder cocaine use over the lifetime (EVERCOC) suggest that two values were substantially different (New Orleans and Portland) and seven values were not substantially different (Dallas, Denver, Indianapolis, Miami, Phoenix, San Antonio, and San Jose). None of the values was indeterminate.

Similarly, for self-reported crack cocaine use over the lifetime (EVERCRK), the distribution is as follows: one value was classified as substantially different (New Orleans), seven values did not show substantial differences (Dallas, Denver, Indianapolis, Phoenix, San Antonio, and San Jose), and one value was statistically indeterminate (Miami).

Figure 5.2. summarizes the results for the drug estimates for cocaine. To reiterate, the following variables were included in this diagram: COC, COC72, CRK72, EVERCOC, and EVERCRK. Of the 45 values for cocaine, eight values (18%) fell into the category "Statistically Different and Not Statistically Equivalent;" 25 values (56%) were classified as either "Statistically Equivalent and Not Statistically Different" or "Statistically Different But Statistically Equivalent;" and 12 values (27%) were neither statistically different nor statistically equivalent. Even though the results for cocaine are not as clear as for marijuana, overall the conclusion would be that the drug use estimates are not substantially different because less than 20% of the drug estimates demonstrated substantial differences.

Third, Tables 5.3. and 5.4. also show the results for the opiate variables. The findings for the urine test results for opiates (OP) show that three values were classified as substantially different (Phoenix, Portland, and San Jose), five values did not show substantial differences Dallas,

Denver, Indianapolis, Phoenix, and Miami), and one value could not be assessed because it was not different and not equivalent (San Antonio).

Figure 5.2. Distribution of Cocaine Values Across the Outcome Categories

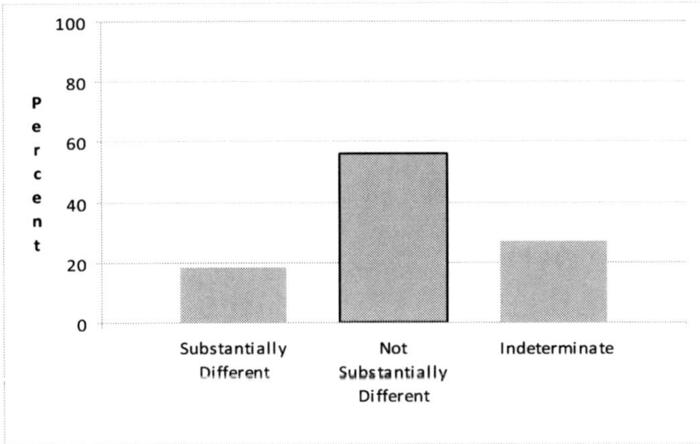

The results for the values for heroin use within the past 72 hours (HER72) demonstrates that none of the values were substantially different, five values were classified as not substantially different (Denver, Miami, New Orleans, Phoenix, and Portland), and four values were statistically indeterminate (Dallas, Indianapolis, San Antonio, and San Jose).

With regard to lifetime heroin use (EVERHER), the analysis demonstrates that two values were substantially different (Phoenix and Portland), six values did not show substantial differences (Denver, Indianapolis, Phoenix, Miami, San Antonio, and San Jose), and one value was statistically indeterminate (Dallas).

Figure 5.3. summarizes the distribution of the values for opiates across the outcome categories. Of the 27 values, five values (19%) showed substantial differences, 16 values (59%) were not substantially different; and six values (22%) could not be statistically determined. Overall, the conclusion is that the drug use estimates for opiates are not substantially different because less than 20% of the values demonstrated substantial differences. After having discussed the overall results and the results for each drug category, the next section analyzes the findings for each site.

Figure 5.3. Distribution of Opiate Values Across the Outcome Categories

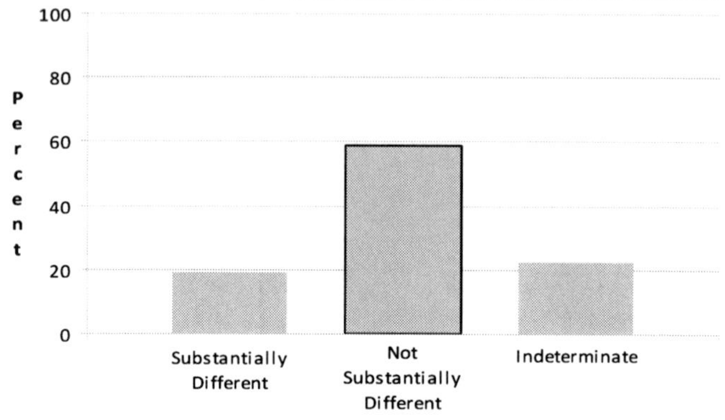

SITE SPECIFIC FINDINGS

Dallas

Table 5.5. presents the results of the equivalence analysis for Dallas. The findings demonstrate that none of the drug use variables fell into the category *different*, that is, none of them were statistically different and not statistically equivalent (D). Further, three variables (27%) were not statistically different in the traditional test and statistically equivalent in the equivalence test, which means that they were classified as *equivalent* (E). These three variables were "Urine Test.

Result for Opiates," "Ever Used Powder Cocaine," and "Ever Used Crack Cocaine." The smaller equivalence z with the higher p-values for these three variables is larger than 1.645, indicating that they are equivalent between DUF and ADAM. The z-values for the traditional test were not larger than 1.96, which means that they were not statistically different. The 90% confidence intervals of these three variables fall within the equivalence interval and include zero. For instance, the equivalence interval for "Ever Used Powder Cocaine" is ±.0648. The 90% confidence interval has a lower limit of -.040 and an upper limit of .028.

Three drug use estimates (27%) were found to be statistically different and statistically equivalent (DE) and thus classified as *different but equivalent*. These variables were "Urine Test Result for

Marijuana," "Marijuana Used within the Past 72 Hours," and "Ever used Marijuana." The z-values for the traditional test were greater than 1.96 and thus displayed a significant difference, but the smaller z-values for the smaller equivalence test were also greater than 1.645 indicating that the confidence intervals overlap with the equivalence interval. In fact, the smaller z-values for the equivalence test were also greater than 1.96 indicating significance at the .025 level.

The remaining five variables were statistically indeterminate because they were *not statistically different and not statistically equivalent* (NDNE). The traditional z was not larger than 1.96 and the smaller equivalence z-value was not larger than 1.645, indicating that neither test was significant and as a result neither of them can be rejected. The confidence limits of these variables were not contained within the equivalence interval. Rather they were partially inside and outside the equivalence margin. The six drug use estimates that were indeterminate were "Urine Test for Cocaine," "Powder Cocaine Used within the Past 72 Hours," "Crack Cocaine Used within the Past 72 Hours," "Heroin Used within the Past 72 Hours," and "Ever Used Heroin."

Figure 5.4. Distribution of Variables across the Outcome Categories in Dallas

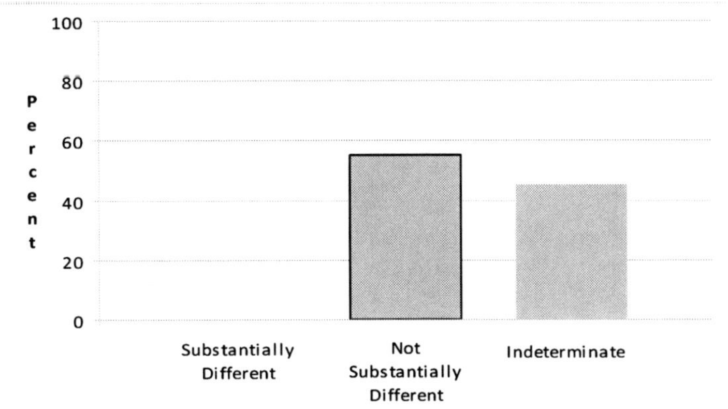

Table 5.5. Equivalence Test DUF and ADAM in Dallas, TX

variable	ADAM 2000/2001		DUF 1997/1998		Difference		Equivalence	Traditional 95% CI				Equivalence 90% CI			
	p1	n1	p2	n2	DIF.	S.E.	Criterion	z	P	LCL	UCL	z	p[a]	LCL	UCL
Urine Test															
Marijuana	0.327	802	0.434	1547	-0.107	0.021	±.0654	-5.141	.000†	-0.148	-0.066	-1.999	.022*	-0.141	-0.073
Cocaine	0.271	802	0.305	1547	-0.034	0.020	±.0542	-1.737	.041†	-0.072	-0.004	1.032	.159	-0.066	-0.002
Opiates	0.042	802	0.035	1547	-0.007	0.021	±.0084	0.825	.203	-0.010	0.024	1.815	.035*	-0.007	0.021
Self-Reported Drug Use Within 72 hours															
Marijuana	0.227	802	0.326	1547	-0.099	0.019	±.0454	-5.212	.000†	-0.136	-0.062	-2.822	.002*	-0.130	-0.068
Cocaine	0.059	802	0.060	1547	-0.001	0.010	±.0118	-0.097	.461	-0.021	0.019	1.051	.159	-0.018	0.016
Crack	0.121	802	0.123	1547	-0.002	0.014	±.0242	-0.141	.444	-0.030	0.026	1.561	.059	-0.025	0.021
Heroine	0.020	802	0.021	1547	-0.001	0.006	±.0040	-0.163	.435	-0.130	0.011	0.488	.316	-0.011	0.009
Ever Used Drug															
Marijuana	0.685	802	0.783	1547	-0.098	0.019	±.1370	-5.035	.000†	-0.136	-0.060	2.004	.023*	-0.130	-0.066
Cocaine	0.324	802	0.330	1547	-0.006	0.020	±.0648	-0.294	.384	-0.046	0.034	2.883	.002*	-0.040	0.028
Crack	0.264	802	0.270	1547	-0.006	0.019	±.0528	-0.312	.378	-0.044	0.032	2.434	.008*	-0.038	0.026
Heroine	0.094	802	0.101	1547	-0.007	0.013	±.0118	-0.545	.293	-0.032	0.018	0.919	.181	-0.028	0.014

Note: Dif. = difference p_1-p_2; S.E. = standard error; CI = confidence interval; LCL = Lower Confidence Limit; UCL = Upper Confidence Limit;
[a]The highest p value of the two one-sided tests has been reported; [b]The equivalence interval was defined as ±20% of the baseline value (ADAM);
†$p \leq .025$ for traditional significance test two-tailed; *$p \leq .05$ for equivalence test, one-tailed

Figure 5.4. summarizes the results for Dallas. Overall, six variables (55%) were not substantially different and five variables (45%) could not be classified because they were statistically indeterminate. In accordance with the decisions rules outlined in the previous chapter, this means that no conclusions can be drawn for Dallas because more than one third, that is, 45%, of the variables could not be statistically assessed.

Denver

Table 5.6. shows the findings for Denver including the 95% confidence interval for the traditional null hypothesis test and 90% confidence limits for the equivalence test and how they fall with regard to the equivalence interval. For Denver about 18% (2) of the variables were substantially *different.* These two variables were "Crack Cocaine Used in the Past 72 Hours," and "Urine Test Result for Cocaine." For these two variables, the z-test for the traditional test was larger than 1.96 and as a result the p-value was smaller than .025 indicating that the difference was statistically significant for the traditional test. With regard to the equivalence test, the smaller z-values with the larger p-values were not larger than 1.645 and as a result the equivalence test was not significant. These equivalence 90% confidence intervals of these two variables were not contained in the equivalence intervals, supporting the finding that they are not equivalent for DUF and ADAM.

Four (36%) of the variables displayed *equivalence*, that is, the smaller equivalence z with the higher p-value was greater than 1.645, indicating statistical equivalence, and the traditional z test was not statistically significant. The four variables that were found to be equivalent were "Urine Test Result for Marijuana," "Urine Test Result for Opiates," Marijuana Used in the Past 72 Hours," and "Ever Used Heroin" As evident in Table 5.6, the 90% equivalence confidence intervals of these variables are contained within the equivalence interval and include zero.

Four (36%) variables fell into the category *different but equivalent*. These four variables were "Heroin Used in the Past 72 Hours," "Ever Used Marijuana," "Ever Used Powder Cocaine," and "Ever Used Crack Cocaine." The traditional z test of these variables implied a statistically significant difference, but the smaller equivalence z was also greater than 1.645. As described previously, this finding indicates that although there is a statistically significant difference this difference is not believed to be substantial, and as a result these drug use estimates can be treated as being comparable.

The remaining variable “Powder Cocaine Used within the Past 72 Hours” was classified as *not different and not equivalent*. The status of this drug use value with regard to equivalence or a substantial difference could not be statistically determined because neither the traditional test nor the equivalence test was statistically significant.

The results for Denver are summarized in Figure 5.5., showing the distribution of the variables across the outcome categories. Overall, two variables were substantially different, eight variables were not substantially different, and one value could not be determined. The findings show that less than 20% of the drug use values were substantially different, which implies that the drug use estimates from DUF and ADAM are not substantially different for the site of Denver.

Figure 5.5. Distribution of Variables across the Outcome Categories in Denver

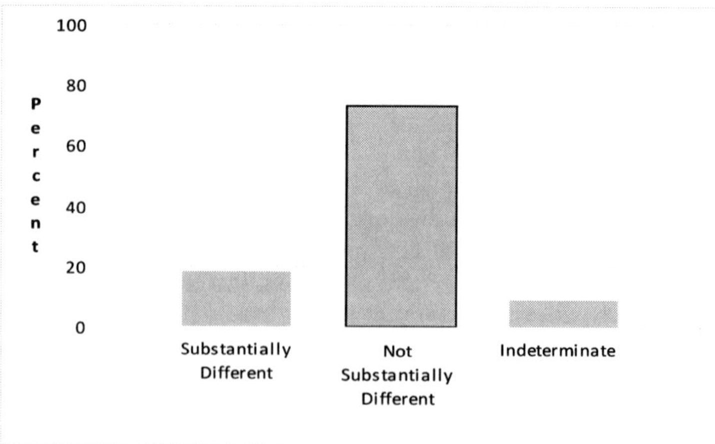

Table 5.6. Equivalence Test DUF and ADAM in Denver, CO

variable	ADAM 2000/2001		DUF 1997/1998		Difference		Equivqlence	Traditional 95% CI				Equivalence 90% CI			
	p1	n1	p2	n2	DIF.	S.E.	Criterion	z	P	LCL	UCL	z	p[a]	LCL	UCL
Urine Test															
Marijuana	0.404	1319	0.413	1908	-0.009	0.018	±.0808	-0.511	.305	-0.043	0.025	4.081	.000*	-0.038	0.020
Cocaine	0.334	1319	0.399	1908	-0.065	0.017	±.0668	-3.789	.000†	-0.099	-0.031	0.105	.460	-0.093	-0.037
Opiates	0.046	1319	0.038	1908	0.008	0.007	±.0092	1.105	.136	-0.006	0.022	2.375	.009*	-0.004	0.020
Self-Reported Drug Use Within 72 hours															
Marijuana	0.303	1319	0.291	1908	0.012	0.016	±.0606	-0.733	.232	-0.020	0.044	4.433	.000*	-0.015	0.039
Cocaine	0.058	1319	0.073	1908	-0.015	0.009	±.0116	-1.711	.044	-0.032	0.002	-0.388	.352	-0.029	-0.001
Crack	0.125	1319	0.152	1908	-0.027	0.012	±.0250	-2.201	.014†	-0.051	-0.003	-0.163	.436	-0.047	-0.007
Heroine	0.026	1319	0.003	1908	0.023	0.005	±.0052	5.047	.000†	-0.014	0.032	6.188	.000*	-0.016	0.030
Ever Used Drug															
Marijuana	0.761	1319	0.855	1908	-0.094	0.014	±.1522	-5.600	.000†	-0.122	-0.066	4.086	.000*	-0.117	-0.071
Cocaine	0.440	1319	0.482	1908	-0.042	0.018	±.0880	-2.356	.009†	-0.077	-0.007	2.581	.006*	-0.070	-0.013
Crack	0.356	1319	0.392	1908	-0.036	0.017	±.0712	-2.083	.019†	-0.070	-0.002	2.037	.023*	-0.064	-0.008
Heroine	0.152	1319	0.156	1908	-0.004	0.007	±.0304	-0.310	.378	-0.029	0.021	2.045	.023*	-0.025	0.017

Note: Dif. = difference p_1-p_2; S.E. = standard error; CI = confidence interval; LCL = Lower Confidence Limit; UCL = Upper Confidence Limit;
[a]The highest p value of the two one-sided tests has been reported; [b]The equivalence interval was defined as ±20% of the baseline value (ADAM);
†$p \leq .025$ for traditional significance test two-tailed; *$p \leq .05$ for equivalence test, one-tailed

Indianapolis

The results for Indianapolis, as presented in Table 5.7., demonstrate that one drug use value (9%) was substantially different. Specifically, the value for "Marijuana Used in the Past 72 Hours" was *statistically different and not statistically equivalent*. The traditional z was -2.194, which equals a p-value of .014 and the equivalence z was 1.293, which equals a p-value of .098. Thus, the variable was significant for the traditional test at the .025 level and not significant for the equivalence test.

Further, four (36%) drug use variables demonstrated statistical *equivalence*. For the variables "Urine Test Results for Cocaine," "Ever Used Marijuana," "Ever Used Powder Cocaine," and "Ever Used Crack Cocaine," the smaller equivalence z was greater than 1.645 and a traditional z that was smaller than 1.96. The 90% confidence interval of these four variables was completely contained within the equivalence interval and included zero.

Additionally, three variables were *different but equivalent*, suggesting trivial differences only. These variables were "Urine Test Results for Marijuana," "Urine Test Results for Opiates," and "Ever Used Heroin." The traditional z was greater than 1.96, implying a significant difference, but the equivalence z of these three variables was also greater than 1.645, suggesting equivalence in the sense that the 90% confidence interval was contained within the equivalence margin. The 90% confidence interval did not include zero, however.

Finally, three variables (27%) were *neither statistically different nor statistically equivalent*. The 90% confidence interval for the variables "Powder Cocaine Used in the Past 72 Hours," "Crack Cocaine Used in the Past 72 Hours," and "Marijuana Used in the Past 72 Hours" fell partially inside and outside the equivalence margin. The traditional z was smaller than 1.96, implying that they were not statistically different, and the equivalence z was smaller than 1.645, suggesting that they were not statistically equivalent either.

Figure 5.6. displays the overall results for Indianapolis. The chart shows that 9% (1) of the variables were substantially different, 64% (7) of the variables were not substantially different, and 27% were statistically indeterminate. These findings support a conclusion of no substantial difference because less than 20% of the variables were substantially different.

Table 5.7. Equivalence Test DUF and ADAM in Indianapolis, IN

variable	ADAM 2000/2001		DUF 1997/1998		Difference		Equivalence	Traditional 95% CI				Equivalence 90% CI			
	p1	n1	p2	n2	DIF.	S.E.	Criterion	z	P	LCL	UCL	z	p[a]	LCL	UCL
Urine Test															
Marijuana	0.487	1362	0.443	1545	0.044	0.019	±.0974	2.375	.009†	0.008	0.080	7.633	.000*	0.014	0.074
Cocaine	0.311	1362	0.325	1545	-0.014	0.017	±.0622	-0.809	.209	-0.048	0.020	2.786	.002*	-0.042	0.014
Opiates	0.051	1362	0.025	1545	0.026	0.007	±.0132	3.630	.000†	0.012	0.040	5.054	.000*	0.014	0.038
Self-Reported Drug Use Within 72 hours															
Marijuana	0.302	1362	0.340	1545	-0.038	0.017	±.0604	-2.194	.014†	-0.072	-0.004	1.293	.098	-0.066	-0.010
Cocaine	0.030	1362	0.032	1545	-0.002	0.006	±.0060	-0.311	.378	-0.015	0.011	0.622	.268	-0.013	0.009
Crack	0.096	1362	0.117	1545	-0.021	0.011	±.0192	-1.838	.033	-0.043	0.001	-0.158	.440	-0.040	-0.002
Heroine	0.013	1362	0.012	1545	0.001	0.004	±.0026	0.242	.405	-0.007	0.009	0.871	.192	-0.006	0.008
Ever Used Drug															
Marijuana	0.811	1362	0.801	1545	0.010	0.015	±.1622	0.681	.248	-0.019	0.039	11.725	.000*	-0.014	0.034
Cocaine	0.318	1362	0.348	1545	-0.030	0.017	±.0636	-1.715	.043	-0.064	0.004	1.920	.027*	-0.059	-0.001
Crack	0.291	1362	0.316	1545	-0.025	0.017	±.0583	-1.465	.071	-0.058	0.008	1.945	.026*	-0.053	0.003
Heroine	0.066	1362	0.112	1545	-0.046	0.010	±.0132	-4.393	.000†	-0.067	-0.025	-3.133	.001*	-0.063	-0.029

Note: Dif. = difference p_1-p_2; S.E. = standard error; CI = confidence interval; LCL = Lower Confidence Limit; UCL = Upper Confidence Limit;

[a]The highest p value of the two one-sided tests has been reported; [b]The equivalence interval was defined as ±20% of the baseline value (ADAM);

†$p \leq .025$ for traditional significance test two-tailed; *$p \leq .05$ for equivalence test, one-tailed

Figure 5.6. Distribution of Variables across the Outcome Categories in Indianapolis

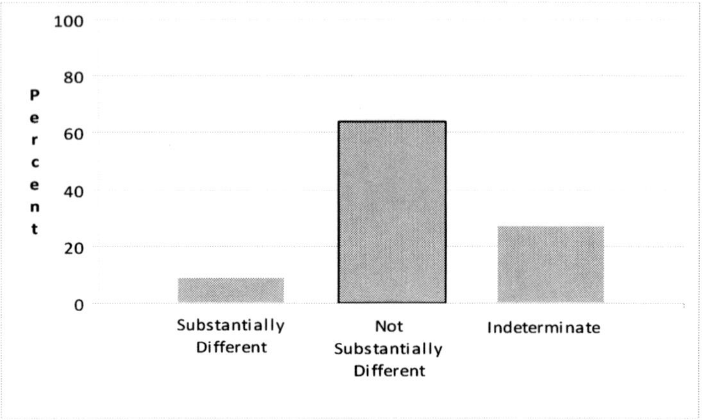

Miami

Table 5.8. shows the outcome of the equivalence analysis for Miami. One value (9%) ("Urine Test Result for Cocaine") was substantially *different* because the traditional z-score was -3.509, which is significant at the .025 level, and the z-score for the equivalence test was -0.008, which is below 1.645 and therefore not statistically significant. The equivalence interval was ±.0898 and the 90% equivalence confidence interval was -.132 for the lower limit and -.048 for the upper limit. Thus, the 90% equivalence interval fell partially inside and outside the equivalence interval and did not include zero.

Five variables (45%) were statistically equivalent and not different and therefore can be considered *equivalent*. These equivalent variables were "Urine Test Result for Marijuana," "Marijuana Used in the Past 72 Hours," "Ever Used Marijuana," "Ever Used Cocaine," and "Ever Used Heroin." The traditional z for the five variables was below 1.96 and as a result not significant and the equivalence z was above 1.645, indicating statistical equivalence. The 90% equivalence confidence interval for these variables was fully contained within the equivalence margin and included zero.

Two variables (18%) fell into the category *different but equivalent*. These variables were "Urine Test Result for Opiates," and "Heroin

Used in the Past 72 Hours." Both variables had traditional z-scores above 1.96, which suggests a significant difference. Their equivalence z-scores were, however, also above 1.645, implying that they were equivalent for DUF and ADAM. Thus, the differences, although statistically significant, are probably small.

Finally, three variables (27%) could not be judged either way because they were neither statistically different nor statistically equivalent. The variables classified into the indeterminate category were "Powder Cocaine Used in the Past 72 Hours," "Crack Cocaine Used in the Past 72 Hours," and "Ever Used Crack". As shown in figure 5.7., only one value (9%) was substantially different and seven variables (64%) did not show substantial differences. Additionally, three variables (27%) had to be classified as indeterminate.

Figure 5.7. Distribution of Variables across the Outcome Categories in Miami

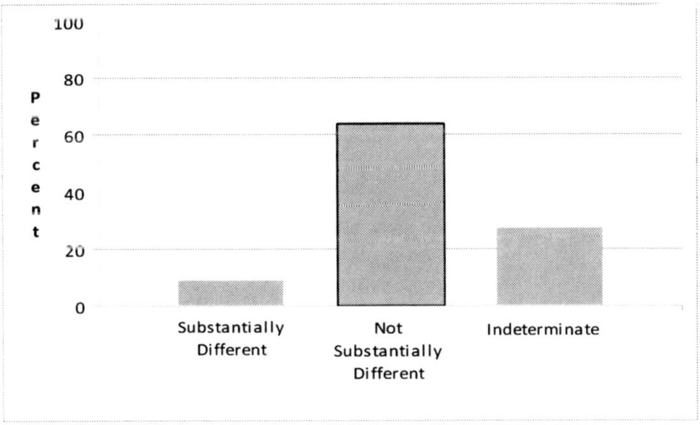

Due to the finding that less than 20% of the drug use estimates were substantially different and less than one-third was statistically indeterminate, the overall conclusion for Miami is that the drug use estimates are not substantially different.

Table 5.8. Equivalence Test DUF and ADAM in Miami, FL

variable	ADAM 2000/2001		DUF 1997/1998		Difference		Equivalence	Traditional 95% CI				Equivalence 90% CI			
	p1	n1	p2	n2	DIF.	S.E.	Criterion	z	P	LCL	UCL	z	p[a]	LCL	UCL
Urine Test															
Marijuana	0.353	535	0.308	1272	0.045	0.024	±.0706	1.846	.033	-0.003	0.093	4.741	.000*	0.005	0.085
Cocaine	0.449	535	0.539	1272	-0.090	0.026	±.0898	-3.509	.000†	-0.140	-0.040	-0.008	.497	-0.132	-0.048
Opiates	0.047	535	0.022	1272	0.025	0.010	±.0094	2.492	.000†	0.005	0.045	3.429	.000*	0.008	0.042
Self-Reported Drug Use Within 72 hours															
Marijuana	0.221	535	0.223	1272	-0.002	0.021	±.0442	-0.093	.463	-0.044	0.040	1.972	.024*	-0.037	0.033
Cocaine	0.097	535	0.113	1272	-0.016	0.016	±.0194	-1.027	.152	-0.047	0.015	0.218	.416	0.042	0.010
Crack	0.121	535	0.149	1272	-0.028	0.017	±.0242	-1.621	.053	-0.062	0.006	-0.220	.413	0.056	0.000
Heroine	0.036	535	0.015	1272	0.021	0.009	±.0072	2.401	.008†	-0.004	0.038	3.225	.000*	0.007	0.035
Ever Used Drug															
Marijuana	0.602	535	0.633	1272	-0.031	0.025	±.1204	-1.235	.108	-0.080	0.018	3.560	.000*	-0.072	0.010
Cocaine	0.370	535	0.384	1272	-0.014	0.025	±.0740	-0.562	.287	-0.063	0.034	2.406	.008*	-0.055	0.027
Crack	0.219	535	0.257	1272	-0.038	0.022	±.0438	-1.735	.040	-0.080	0.004	0.268	.397	-0.074	-0.002
Heroine	0.090	535	0.076	1272	-0.014	0.014	±.0180	0.970	.166	-0.014	-0.042	2.217	.014*	-0.010	0.038

Note: Dif. = difference p_1-p_2; S.E. = standard error; CI = confidence interval; LCL = Lower Confidence Limit; UCL = Upper Confidence Limit;
[a]The highest p value of the two one-sided tests has been reported; [b]The equivalence interval was defined as ±20% of the baseline value (ADAM);
†$p \leq .025$ for traditional significance test two-tailed; *$p \leq .05$ for equivalence test, one-tailed

New Orleans

Table 5.9. displays the findings for New Orleans. The drug use estimates for “Ever Used Powder Cocaine” and “Ever Used Crack Cocaine” were categorized as *different*. The 90% equivalence confidence interval for both variables fell partially inside and partially outside the equivalence interval and did not include zero. Specifically, the equivalence interval for “Ever Used Powder Cocaine” was ±.0558. The lower limit of the 90% equivalence confidence interval was -.090 and the upper limit was -.036. Similarly, the equivalence interval for “Ever Used Crack Cocaine” was ±.0512. The lower limit of the 90% equivalence confidence interval was -.083 and the upper limit was -.029. Also, the traditional z-value for “Ever Used Powder Cocaine” was -3.781 and for “Ever Used Crack Cocaine” was -3.449, which is associated with a significant p-value of .000 for both variables. The smaller equivalence z-value for the two variables was below 1.645, indicating that there is no equivalence. Thus, these two variables were classified as substantially different.

Four variables (36%) demonstrated *equivalence* for DUF and ADAM, that is, the drug use estimates for ”Marijuana Used in the Past 72 Hours,” ”Heroin Used in the Past 72 Hours,” “Ever Used Marijuana,“ and “Ever Used Heroin” were statistically equivalent and not statistically different. The 90% equivalence confidence interval of these variables fell within the equivalence margin and included zero. Also, the traditional z-value was below 1.96, suggesting that no statistically significant differences existed between the drug use estimates for DUF and ADAM and the equivalence z-value was greater than 1.645, indicating that the drug use estimates were statistically equivalent.

Additionally, five drug use variables (45%) showed trivial differences only because they fell into the category *different but equivalent*. These five variables were “Urine Test Results for Marijuana,” “Urine Test Results for Cocaine,” “Urine Test Results for Opiates,” “Powder Cocaine Used in the Past 72 Hours,” and “Crack Cocaine Used in the Past 72 Hours.” The traditional z-value for these variables was above 1.96, indicative of a statistically significant difference. At the same time, the smaller equivalence z was also statistically significant demonstrating that the drug use estimates of these five variables were equivalent for DUF and ADAM. Accordingly, the five variables were classified as not substantially different.

Table 5.9. Equivalence Test DUF and ADAM in New Orleans, LA

variable	ADAM 2000/2001		DUF 1997/1998		Difference		Equivalence	Traditional 95% CI				Equivalence 90% CI			
	p1	n1	p2	n2	DIF.	S.E.	Criterion	z	P	LCL	UCL	z	p[a]	LCL	UCL
Urine Test															
Marijuana	0.454	1236	0.382	1959	0.072	0.018	±.0908	4.018	.000†	0.037	0.107	9.086	.000*	0.043	0.101
Cocaine	0.350	1236	0.457	1959	-0.107	0.018	±.0700	-6.070	.000†	-0.142	-0.072	-2.099	.018	-0.136	-0.078
Opiates	0.151	1236	0.118	1959	0.033	0.013	±.0302	2.635	.004†	0.008	0.058	5.046	.000*	0.012	0.054
Self-Reported Drug Use Within 72 hours															
Marijuana	0.350	1236	0.331	1959	0.019	0.017	±.0700	1.102	.163	-0.015	0.053	5.163	.000*	-0.009	0.047
Cocaine	0.057	1236	0.085	1959	-0.028	0.009	±.0114	-3.070	.001†	-0.046	-0.010	-1.820	.034	-0.043	-0.013
Crack	0.121	1236	0.173	1959	-0.052	0.013	±.0242	-4.123	.000†	-0.077	-0.027	-2.204	.014*	-0.073	-0.031
Heroine	0.112	1236	0.092	1959	0.020	0.011	±.0224	1.803	.036	-0.002	0.042	3.821	.000*	0.002	0.038
Ever Used Drug															
Marijuana	0.758	1236	0.773	1959	-0.015	0.015	±.1516	-0.972	.166	-0.045	0.015	8.855	.000*	-0.040	0.010
Cocaine	0.279	1236	0.342	1959	-0.063	0.017	±.0558	-3.781	.000†	-0.096	-0.030	-0.432	.333	-0.090	-0.036
Crack	0.256	1236	0.312	1959	-0.056	0.016	±.0512	-3.449	.000†	-0.088	-0.024	-0.296	.394	-0.083	-0.029
Heroine	0.206	1236	0.208	1959	-0.002	0.015	±.0412	-0.136	.444	-0.031	0.027	2.665	.004*	-0.026	0.022

Note: Dif. = difference p_1-p_2; S.E. = standard error; CI = confidence interval; LCL = Lower Confidence Limit; UCL = Upper Confidence Limit;
[a]The highest p value of the two one-sided tests has been reported; [b]The equivalence interval was defined as ±20% of the baseline value (ADAM);
†$p \leq .025$ for traditional significance test two-tailed; *$p \leq .05$ for equivalence test, one-tailed

All of the drug use variables were classified as either substantially different, equivalent, or equivalent but different. Figure 5.8. shows the distribution of the drug use estimates across the outcomes categories. About 18% of the drug use estimates were substantially different, and 82% were classified as not substantially different. Based on these results, the overall finding for New Orleans is that there is no substantial difference between the drug estimates for DUF and ADAM because less than 20% of the drug use variables demonstrated substantial differences.

Figure 5.8. Distribution of Variables Across the Outcome Categories in New Orleans

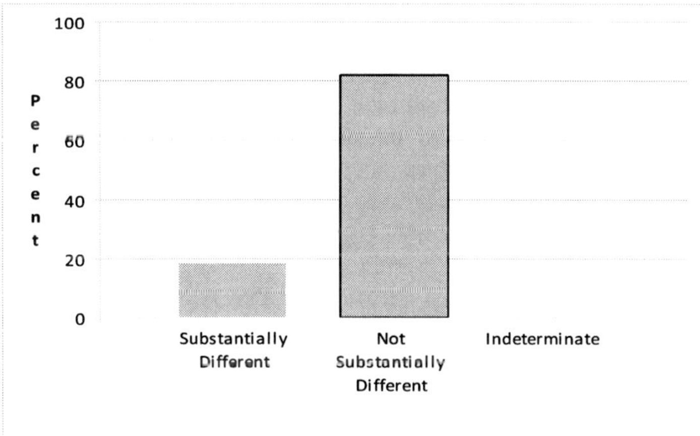

Phoenix

The results for Phoenix, as presented in Table 5.10., are somewhat different than the patterns of the previous five sites. The main difference is that Phoenix has the largest amount of drug use estimates that are substantially *different*. Specifically, in Phoenix four variables (36%) fell into the category statistically different and not statistically equivalent. These were "Urine Test Result for Cocaine," and "Urine Test Result for Opiates," "Cocaine Used in the Past 72 Hours," and "Ever Used Heroin." The traditional z-values for these drug estimates were greater 1.96 and as a result significant at the .025 level. Also, the smaller equivalence z-values were not significant, suggesting that equivalence does not exist. The 90% equivalence confidence intervals for these four variables were not contained within the equivalence margin. Rather, they were partially inside and partially outside the margin and did not include zero.

Three drug use variables (27%) fell into the category statistically equivalent and not statistically different and were classified as *equivalent*. These three variables were "Ever Used Marijuana," "Ever Used Powder Cocaine," and "Ever Used Crack Cocaine." The 90% equivalence confidence intervals for all three variables fell completely inside the equivalence margin, and the equivalence z-values were statistically significant, indicating equivalence. Finally, the traditional z-values were not significant, suggesting that no statistically significant differences existed.

Another three drug use variables—"Urine Test Result for Marijuana," "Marijuana Used in the Past 72 Hours," and "Heroin Used in the Past 72 Hours" —fell into the category *different but equivalent.* These three variables showed statistically significant differences but they were also statistically equivalent and thus categorized as not substantially different.

One drug use estimate ("Crack Cocaine Used in the Past 72 Hours") was classified as indeterminate because the drug use variables for DUF and ADAM were not statistically different and not statistically equivalent. The 90% equivalence confidence interval for this value fell partially inside and partially outside the equivalence margin and did not contain zero. The equivalence margin was ±.0252. The lower limit of the confidence interval was -.038 and the upper limit was -.002.

Figure 5.9. summarizes the results for Phoenix. These findings suggest that there is some evidence for substantial differences between the drug estimates for DUF and ADAM. Specifically, 36% of the drug use estimates indicated substantial differences. Only about 55% of the drug use estimates showed no substantial differences and one variable could not be assessed. In accordance with the decision rules outlined earlier the results for Phoenix are interpreted as substantially different because more than 20% of the drug use estimates were classified as *different*.

Figure 5.9. Distribution of Variables across the Outcome Categories in Phoenix

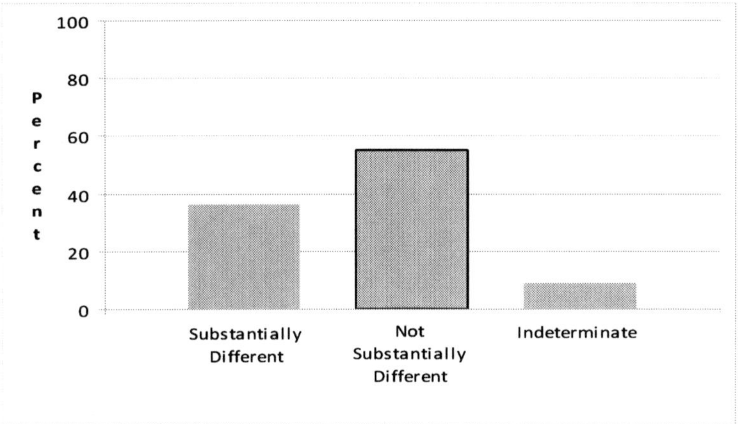

Table 5.10. Equivalence Test DUF and ADAM in Phoenix, AZ

Variable	ADAM 2000/2001		DUF 1997/1998		Difference		Equivalence	Traditional 95% CI				Equivalence 90% CI			
	p1	n1	p2	n2	DIF.	S.E.	Criterion	z	P	LCL	UCL	z	p[a]	LCL	UCL
Urine Test															
Marijuana	0.359	2850	0.311	1611	0.048	0.015	±.0718	3.283	.000†	0.019	0.077	8.194	.000*	0.024	0.072
Cocaine	0.276	2850	0.318	1611	-0.042	0.014	±.0552	-2.935	.002†	-0.070	-0.014	0.923	.179	-0.066	-0.018
Opiates	0.062	2850	0.079	1611	-0.017	0.008	±.0124	-2.099	.018†	-0.033	-0.001	-0.568	.285	-0.030	-0.004
Self-Reported Drug Use Within 72 hours															
Marijuana	0.261	2850	0.220	1611	0.041	0.013	±.0552	3.106	.000†	0.015	0.067	7.062	.000*	0.019	0.063
Cocaine	0.050	2850	0.066	1611	-0.016	0.007	±.0100	-2.159	.015†	-0.031	-0.001	-0.810	.209	-0.028	-0.004
Crack	0.126	2850	0.146	1611	-0.020	0.011	±.0252	-1.857	.032	-0.041	0.001	0.483	.319	-0.038	-0.002
Heroine	0.045	2850	0.073	1611	-0.028	0.008	±.0090	-3.706	.000†	-0.043	-0.013	-2.515	.006*	-0.040	-0.016
Ever Used Drug															
Marijuana	0.807	2850	0.814	1611	-0.007	0.012	±.1614	-0.574	.283	-0.031	0.017	12.665	.000*	-0.027	0.013
Cocaine	0.508	2850	0.498	1611	-0.010	0.016	±.1016	0.642	.261	-0.021	0.041	7.161	.000*	-0.016	0.036
Crack	0.395	2850	0.416	1611	-0.021	0.015	±.0790	-1.371	.085	-0.051	0.009	3.786	.000*	-0.046	0.004
Heroine	0.175	2850	0.219	1611	-0.044	0.013	±.0350	-3.514	.000†	-0.069	0.019	-0.719	.764	-0.065	-0.023

Note: Dif. = difference p_1-p_2; S.E. = standard error; CI = confidence interval; LCL = Lower Confidence Limit; UCL = Upper Confidence Limit;

[a]The highest p value of the two one-sided tests has been reported; [b]The equivalence interval was defined as ±20% of the baseline value (ADAM);

†$p \leq .025$ for traditional significance test two-tailed; *$p \leq .05$ for equivalence test, one-tailed

Portland

The results for Portland, as presented in Table 5.11., demonstrate that three (27%) of the drug use variables were classified as *different*. These variables were "Urine Test Results for Opiates," "Ever Used Powder Cocaine," and "Ever Used Heroin." All three variables had traditional z-scores above 1.96, and smaller equivalence z-values fell below 1.645. The 90% equivalence confidence interval of these variables was not contained within the equivalence margin and did not include zero.

Also, three drug use variables (27%) fell into the category *equivalent*; including "Urine Test Results for Marijuana" "Marijuana Used Within the Past 72 Hours," and "Ever Used Crack Cocaine." For these three variables, the 90% equivalence confidence interval fell within the equivalence margin and included zero. The traditional z was not statistically significant indicating that there was no significant difference, and the smaller equivalence z was statistically significant, suggesting that the drug use estimates for these variables were equivalent for DUF and ADAM.

Additionally, five (45%) of the drug use estimates were *different but equivalent*. These estimates were "Urine Test Result for Cocaine," "Powder Cocaine Used Within the Past 72 Hours," "Crack Cocaine Used Within the Past 72 Hours," "Heroin Used Within the Past 72 Hours," and "Ever Used Marijuana." For these variables the traditional z was statistically significant, indicating that there were significant differences for the drug estimates of DUF and ADAM. At the same time, the smaller equivalence z was statistically significant and the 90% equivalence confidence interval fell within the equivalence margin, suggesting that the differences between the drug estimates for DUF and ADAM were only slightly different.

None of the drug use variables were statistically indeterminate, that is, all variables were classified as either substantially *different*, *equivalent*, or *equivalent but different*. Figure 5.10. shows the findings for Portland. Overall, more than 20% of the drug use estimates are substantially different for DUF and ADAM and as a result the conclusions for Portland are that there are substantial differences.

Table 5.11. Equivalence Test DUF and ADAM in Portland, OR

Variable	ADAM 2000/2001		DUF 1997/1998		Difference		Equivalence	Traditional 95% CI				Equivalence 90% CI			
	p1	n1	p2	n2	DIF.	S.E.	Criterion	z	P	LCL	UCL	z	p[a]	LCL	UCL
Urine Test															
Marijuana	0.354	1295	0.375	1401	-0.021	0.019	±.0708	-1.132	.129	0.057	0.015	2.685	.000*	-0.052	0.010
Cocaine	0.236	1295	0.327	1401	-0.091	0.017	±.0472	-5.286	.000†	-0.125	-0.057	-2.544	.005	-0.119	-0.063
Opiates	0.114	1295	0.148	1401	-0.034	0.013	±.0228	-2.623	.004†	-0.059	-0.009	-0.864	.194	-0.055	-0.013
Self-Reported Drug Use Within 72 hours															
Marijuana	0.247	1295	0.248	1401	-0.001	0.017	±.0494	-0.060	.476	0.032	0.032	2.909	.002*	-0.028	0.026
Cocaine	0.050	1295	0.079	1401	-0.029	0.009	±.0100	-3.081	.001†	-0.011	-0.011	-2.018	.022*	-0.044	-0.014
Crack	0.099	1295	0.140	1401	-0.041	0.012	±.0198	-3.295	.000†	-0.017	-0.017	-1.704	.044*	-0.061	-0.021
Heroine	0.065	1295	0.110	1401	-0.045	0.011	±.0130	-4.164	.000†	-0.024	-0.024	-2.961	.002*	-0.063	-0.027
Ever Used Drug															
Marijuana	0.841	1295	0.874	1401	-0.033	0.013	±.1682	-2.447	.007†	-0.059	-0.007	10.025	.000*	-0.055	-0.011
Cocaine	0.465	1295	0.541	1401	-0.076	0.019	±.0930	-3.955	.000†	-0.114	-0.038	0.885	.812	-0.108	-0.044
Crack	0.411	1295	0.445	1401	-0.034	0.019	±.0822	-1.784	.037	-0.071	0.003	2.529	.006*	-0.065	-0.003
Heroine	0.242	1295	0.310	1401	-0.068	0.017	±.0484	-3.964	.000†	-0.102	-0.034	-1.142	.127	-0.096	-0.040

Note: Dif. = difference p_1-p_2; S.E. = standard error; CI = confidence interval; LCL = Lower Confidence Limit; UCL = Upper Confidence Limit;
[a]The highest p value of the two one-sided tests has been reported; [b]The equivalence interval was defined as ±20% of the baseline value (ADAM);
†$p \leq .025$ for traditional significance test two-tailed; *$p \leq .05$ for equivalence test, one-tailed

Figure 5.10. Distribution of Variables Across the Outcome Categories in Portland

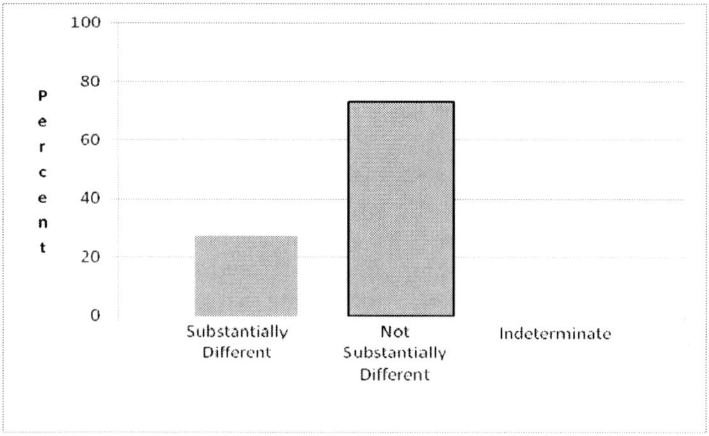

San Antonio

The results for San Antonio are shown in Table 5.12. None of the drug use variables imply a substantial *difference*, that is, none of them were statistically different and not statistically equivalent. Six variables (55%) were classified as *equivalent* because the smaller equivalence z-values of these drug estimates were statistically significant, indicating that they were equivalent for DUF and ADAM. Contrary, the traditional z-values were not statistically significant, providing evidence that there were no significant differences for these drug estimates for DUF and ADAM. These variables were the drug use estimates for "Urine Test Result for Marijuana," "Urine Test Result for Cocaine," "Marijuana Used in the Past 72 Hours," "Crack Cocaine Used in the Past 72 Hours," "Ever Used Marijuana," and "Ever Used Heroin." The 90% equivalence confidence interval for these six variables was completely contained within the equivalence margin and included zero.

Two variables (18%) fell into the category *different but equivalent*. This category included the variables "Ever Used Marijuana" and "Ever Used Crack Cocaine." The traditional z-value for these two variables showed a statistically significant difference between the drug estimates but the smaller equivalence z was also statistically significant, indicating that the difference was not substantial. The 90% equivalence confidence interval for these two variables fell completely within the equivalence margin but did not include zero.

Finally, three variables (27%) fell into the category *not different and not equivalent*, and as a result were classified as statistically indeterminate. These variables were "Urine Test Analysis for Opiates," "Powder Cocaine Used in the Past 72 Hours," and "Heroin Used in the Past 72 Hours."

Figure 5.11. displays the summary results for San Antonio. These results indicate that 73% of the drug use estimates were not substantially different for DUF and ADAM and that none of drug use estimates suggested substantial differences. Three variables could not be judged either way. Overall, the results can be said to be not substantially different.

Figure 5.11. Distribution of Variables Across the Outcome Categories in San Antonio

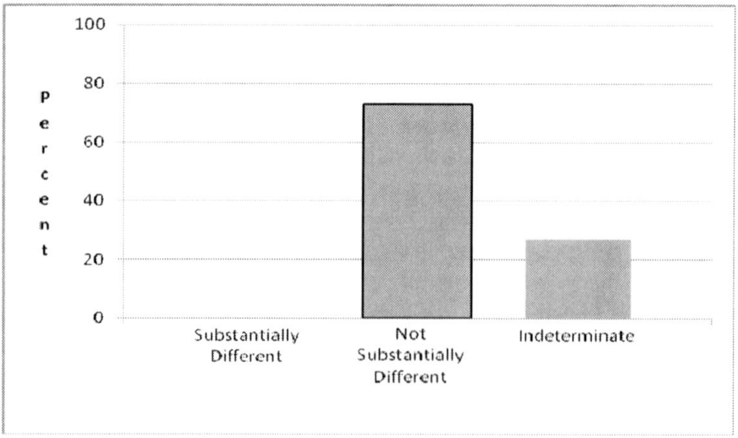

Table 5.12. Equivalence Test DUF and ADAM in San Antonio, TX

Variable	ADAM 2000/2001		DUF 1997/1998		Difference		Equivalence	Traditional 95% CI				Equivalence 90% CI			
	p1	n1	p2	n2	DIF.	S.E.	Criterion	z	P	LCL	UCL	z	p[a]	LCL	UCL
Urine Test															
Marijuana	0.365	1122	0.377	1840	-0.012	0.018	±.0730	-0.656	.256	-0.048	0.024	3.337	.000*	-0.042	0.018
Cocaine	0.253	1122	0.266	1840	-0.013	0.017	±.0506	-0.785	.216	-0.045	0.019	2.269	.011*	-0.040	0.014
Opiates	0.091	1122	0.099	1840	-0.008	0.011	±.0182	-0.724	.135	-0.030	0.014	0.923	.179	-0.026	0.010
Self-Reported Drug Use Within 72 hours															
Marijuana	0.263	1122	0.250	1840	0.013	0.017	±.0526	-0.784	.218	-0.019	0.045	3.958	.000*	-0.014	0.040
Cocaine	0.085	1122	0.089	1840	-0.004	0.011	±.0170	-0.376	.353	-0.025	0.017	1.221	.111	-0.022	0.014
Crack	0.033	1122	0.028	1840	0.005	0.007	±.0066	0.760	.224	-0.008	0.018	1.764	.039*	-0.006	0.016
Heroine	0.058	1122	0.051	1840	0.007	0.009	±.0116	0.808	.212	-0.010	0.024	2.148	.016*	-0.007	0.021
Ever Used Drug															
Marijuana	0.700	1122	0.639	1840	0.061	0.018	±.1400	3.450	.000†	0.026	0.096	11.370	.000*	0.032	0.090
Cocaine	0.356	1122	0.323	1840	0.033	0.018	±.0720	1.836	.034	-0.002	0.068	5.796	.000*	0.003	0.063
Crack	0.148	1122	0.122	1840	0.026	0.013	±.0296	1.991	.023	-0.000	0.052	4.257	.000*	0.005	0.047
Heroine	0.135	1122	0.136	1840	-0.001	0.013	±.0270	-0.077	.469	-0.026	0.024	2.006	.023*	-0.022	0.020

Note: Dif. = difference p_1-p_2; S.E. = standard error; CI = confidence interval; LCL = Lower Confidence Limit; UCL = Upper Confidence Limit;

[a]The highest p value of the two one-sided tests has been reported; [b]The equivalence interval was defined as ±20% of the baseline value (ADAM);

†$p \leq .025$ for traditional significance test two-tailed; *$p \leq .05$ for equivalence test, one-tailed

San Jose

Finally, the results for San Jose are shown in Table 5.13. Only one variable (9%) proved to be *different* ("Urine Test Result for Opiates"). The traditional z-value showed a statistically significant difference for the drug estimates for DUF and ADAM. The smaller equivalence z-value fell below 1.645, indicating that equivalence did not exist. The equivalence margin for this variable was ±.0070, and the 90% equivalence confidence interval was -.030 for the lower limit and -.004 for the upper limit. That shows that the confidence interval falls partially inside and outside the equivalence margin and does not include zero.

Three variables (27%) fell into the category *equivalent*, including "Urine Test Result for Cocaine," "Crack Used within the Past 72 Hours," and "Ever Used Heroin." The traditional z-value did not demonstrate a statistically significant difference and the smaller equivalence z-value did suggest that there is equivalence between the drug estimates. The 90% equivalence confidence interval fell fully within the equivalence margin for these drug estimates and included zero.

Additionally, five variables (45%) were categorized as statistically *different and equivalent*, suggesting the presence of only trivial differences. These five variables included "Urine Test Result for Marijuana," "Marijuana Used within the Past 72 Hours," "Ever Used Marijuana," "Ever Used Powder Cocaine," and "Ever Used Crack Cocaine." These five variables showed statistically significant differences but they also showed statistically significant equivalence.

Two variables (18%) could not be statistically determined. They were *neither statistically different nor equivalent*. These two variables were "Cocaine Used in the Past 72 Hours" and "Heroin Used in the Past 72 Hours." The 90% confidence interval for these two variables fell partially inside and partially outside the equivalence margin but included zero.

Figure 5.12. summarizes the results for San Jose. Overall, 71% of the variables demonstrated no substantial differences. Only one variable was substantially different and two variables could not be statistically assessed. Consistent with the decision criteria the site of San Jose can be said to have drug estimates for DUF and ADAM that are not substantially different.

Table 5.13. Equivalence Test DUF and ADAM in San Jose, CA

Variable	ADAM 2000/2001		DUF 1997/1998		Difference		Equivalence	Traditional 95% CI				Equivalence 90% CI			
	p1	n1	p2	n2	DIF.	S.E.	Criterion	z	P	LCL	UCL	z	p[a]	LCL	UCL
Urine Test															
Marijuana	0.351	1248	0.275	1319	0.076	0.018	±.0702	4.160	.000†	0.040	0.112	8.003	.000*	0.046	0.106
Cocaine	0.120	1248	0.118	1319	0.002	0.013	±.0240	0.156	.440	-0.023	0.027	2.033	.023*	-0.019	0.023
Opiates	0.035	1248	0.052	1319	-0.017	-0.008	±.0070	-2.118	.017†	-0.033	-0.001	-1.246	.106	-0.030	-0.004
Self-Reported Drug Use Within 72 hours															
Marijuana	0.263	1248	0.189	1319	0.074	0.016	±.0526	4.491	.000†	0.042	0.106	7.683	.000*	0.047	0.101
Cocaine	0.032	1248	0.040	1319	-0.008	0.007	±.0024	-1.089	.138	-0.022	0.006	-0.218	.414	-0.020	0.004
Crack	0.042	1248	0.034	1319	0.008	0.008	±.0084	1.058	.159	-0.007	0.023	2.170	.015*	-0.004	0.020
Heroine	0.018	1248	0.028	1319	-0.010	0.006	±.0036	-1.695	.045	-0.022	0.002	-1.085	.139	-0.002	0.000
Ever Used Drug															
Marijuana	0.714	1248	0.670	1319	0.044	0.018	±.1428	2.418	.008†	0.008	0.080	10.264	.000*	0.014	0.074
Cocaine	0.443	1248	0.379	1319	0.064	0.019	±.0886	3.300	.000†	0.026	0.102	7.868	.000*	0.032	0.096
Crack	0.276	1248	0.207	1319	0.069	0.017	±.0552	4.090	.000†	0.036	0.102	7.363	.000*	0.041	0.097
Heroine	0.118	1248	0.107	1319	-0.011	0.012	±.0236	0.881	.189	-0.013	0.035	2.772	.003*	-0.010	0.032

Note: Dif. = difference p_1-p_2; S.E. = standard error; CI = confidence interval; LCL = Lower Confidence Limit; UCL = Upper Confidence Limit; [a]The highest p value of the two one-sided tests has been reported; [b]The equivalence interval was defined as ±20% of the baseline value (ADAM); †$p \leq .025$ for traditional significance test two-tailed; *$p \leq .05$ for equivalence test, one-tailed

Figure 5.12. Distribution of Variables Across the Outcome Categories in San Jose

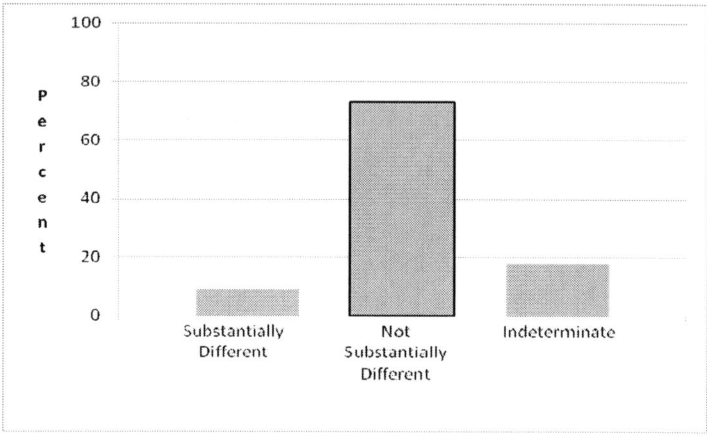

THE IMPACT OF DIFFERENT ALPHA LEVELS AND THE INCLUSION/EXCLUSION OF INDETERMINATE VALUES

Table 5.14 shows that changes in the outcomes due to using different alpha levels and manipulating the inclusion and exclusion of the values found to be statistically indeterminate. The data shows that these manipulations can make a difference with regard to the findings. The analysis of all possible combinations is beyond the scope of this research, and as a result, only four possibilities will be discussed. These four possibilities are shown in Table 5. Column one shows the results for the original analysis using a .05 alpha level for the traditional null hypothesis test and including the values that were statistically indeterminate in the analysis. Column two presents the findings if the indeterminate values would have been excluded from the calculation. The alpha level was kept at .05 to determine changes only due to the exclusion of certain values. Column three shows the results if a .01 alpha level would have been used instead of a .05 level. The indeterminate values were included in the analysis (as in the original analysis) to determine changes due solely to the more conservative alpha level. Column four illustrates the findings for a change in the alpha level to .01 and the exclusion of indeterminate values.

With regard to the indeterminate values, it could be argued that the results of the analysis might be different if the values that were indeterminate would be excluded. For instance, assume that one site had two values that showed substantial differences, six values that were similar, and three values that could not be determined as either different or similar because neither the traditional nor the equivalence test were significant. In the original analyses the total number of values would be eleven. Out of these eleven values, two values (18%) were substantially different, six values (55%) were similar, and three values (27%) were indeterminate. The conclusion would be that there are no substantial differences because less than 20% of the values demonstrated substantial differences.

This conclusion might be different if the three values that could not be statistically determined would be excluded; that is, the total number of values that can be evaluated is eight rather than eleven. For the current example this means that two out of eight values would be substantially different, which equals 25%; and six values are similar, which equals 75%. The conclusion would be that this site demonstrates substantial differences because more than 20% of the values demonstrated substantial differences. Thus, the question is whether the findings of the current study would be different if the indeterminate values would be excluded.

Column two presents the results for this analysis. Column one is the reference category presenting the results from the original analysis. Column two suggests that excluding the indeterminate values would lead to differences in the results for two sites. In Dallas, the results would now indicate that there are no substantial differences. Specifically, in Dallas, none of the variables were categorized as substantially different, three variables were equivalent, three variables were different but equivalent, and five variables could not be statistically assessed. Thus, the new number of total variables would be six (11 – 5) and the conclusion would be that since no variable showed substantial differences the data for Dallas can be said to be similar and not substantially different. This is different from the original conclusion in that Dallas was categorized originally as indeterminate because more than one third of the values could not be statistically judged to be different or similar.

The second site that would be categorized differently if the indeterminate values would be excluded is Denver. In Denver one variable was indeterminate, which results in a new total of variables. The new total would be 10 (11 – 1). In Denver two values were categorized as substantially different, which is exactly 20%. Thus, Denver could be said to show substantial differences across the DUF and ADAM data.

None of the remaining seven sites would have been categorized differently. This means that overall, if the indeterminate values would be excluded from the analysis, six sites would be said to be similar and three sites would be said to be substantially different (as shown in column two in Table 5). The conclusions from the original analysis were that one site could not be categorized as either similar or different because too many values were indeterminate, two sites were substantially different, and six sites were similar.

With regard to the alpha level, it could be argued that the alpha-level used for the analysis should be .01 instead of .05 because the sample sizes (N) used in the current analysis are large. Due to the fact that N influences the outcome of the traditional significance test, it would be reasonable to use the smaller alpha level to reduce the Type I error. A Type I error leads us to reject the null hypothesis of "no difference" and conclude that there is a difference when in fact the null hypothesis of "no difference" is true. At the .05 level there is a 5% chance that the null hypothesis of "no difference" is correct given these data. Accordingly, at the .01 level there is a 1% chance that the null hypothesis of "no difference" is correct given these data. Thus, the .01 level is the more conservative level. The greater the sample size, the smaller or more conservative the alpha level should be in order to avoid a Type I Error.

If the alpha level is more conservative, however, the chances of making the Type II Error increase, that is, researchers will erroneously accept the null hypothesis of "no difference" when in fact there is a significant difference and the null hypothesis should be rejected. Researchers have to decide which alpha level is suited best for their data. For the current study, the sample sizes vary between 535 and 2,850 to ADAM and 1,272 to 1,959 for DUF (as shown in Table 3.4 in Chapter Three). Thus, according to Cohen's Power Primer, an alpha level .01 could be used instead of the .05 level. Column Three in Table 5 presents the results for the analysis if an alpha level of .01 would have been used in the current analysis instead of .05 (as shown in Column One). Seven of the nine sites would have been classified into the same category at the .01 and the .05 alpha level. These seven sites are Dallas ("Unknown"), Denver ("Similar"), Miami ("Similar"), New

Orleans ("Similar"), Portland ("Different"), San Antonio ("Similar"), and San Jose ("Similar").

Two sites would have been categorized differently. First, Indianapolis was originally classified as "Similar." Using the .01 alpha level the results for Indianapolis would have changed and it would have been classified as "unknown." Specifically, in Indianapolis, marijuana used within the last 72 hours was statistically different at the .05 level but not at the .01 level. There was no statistical equivalence for this variable, and thus the variable was classified as indeterminate at the .01 level, increasing the number of indeterminate values to four, which is 36%. Thus, at the .01 alpha level, more than one third of the variables were statistically indeterminate and as a result Indianapolis cannot be said to be either different or similar. No judgment can be made for this site and it would be categorized as "unknown."

Second, Phoenix would have been classified as "similar" at the .01 level whereas it was "different" at the .05 level. This change occurred because at the .05 level four variables were statistically different and not equivalent, and as a result more than 20% of the variables were different leading to a classification of substantially different. At the .01 level, only two variables were statistically different and not equivalent, which is 18%, and as a result Indianapolis would have been categorized as "similar."

Finally, the question is what would happen if the variables that are statistically indeterminate would be excluded from the calculation and an alpha level .01 would be used instead of .05. Column four shows the findings for this analysis. Again, the reference category is Column one (original analysis). Comparing Column four and Column one shows that eight of the nine sides would have been categorized into the same category. Only one side (Dallas) would have been judged differently, that is Dallas would have been classified as "similar" instead of "unknown".

The results presented in Table 5 suggest that the classification for five sides (Miami, New Orleans, Portland, San Antonio, and San Jose) did not change regardless of the intervention. Stated differently, the findings for these five sides were the same in the original analysis, at the .01 level, and excluding the values that were indeterminate. This suggests that the results for these sites are very stable and the conclusions can be drawn with confidence.

For four siTes, however, some changes took place when using a different alpha level and excluding the values that were indeterminate from the analysis. Dallas showed the most changes. In the original analysis (Column One) and at the .01 alpha level (Column Three) Dallas was categorized as “unknown.” This classification would have changed if the indeterminate values would have been excluded (Column Two) and if the analysis would have employed a .01 alpha level and excluded the indeterminate values (Column Four). In this case Dallas would have been judged to be “similar.”

The findings for Indianapolis would have changed from being “similar” to “unknown” if the .01 level would have been used with everything else being the same (Column Three). At all other conditions, Indianapolis was classified as “similar.” The site Phoenix would be classified as “different” under all conditions except when using the .01 alpha level instead of .05 (Column Three). Denver was classified as “similar” in all circumstances except when the indeterminate variables were excluded from the analysis (Column Two), in which situation Denver would have been classified as “different.”

Overall, it appears that the results of the current analysis are very consistent for some sites, but also show some differences across different conditions. The consistency of the findings is especially obvious in column four as compared to column one, where only one site (Dallas) would have been classified differently, as “similar” rather than “unknown,” if both changes (using a .01 alpha level and excluding indeterminate values) would have been made. This would strengthen the conclusion that the drug estimates contained in the DUF and ADAM data are similar. Further, the data suggests that the drug use estimates contained in the DUF and ADAM data show no differences for five sites regardless of the intervention. These sites are Miami, New Orleans, Portland, San Jose, and San Antonio. There are, however, also some differences. Classification would have changed for four sites: Dallas, Denver, Indianapolis, and Phoenix. There is no consistency with regard to the direction of change. The classification of Dallas and Indianapolis varied between “similar” and “unknown,” Denver and Phoenix varied between “similar” and “different.” The implications of these changes will be discussed in the final chapter.

Certainly, the changes of the alpha level and the inclusion and exclusion criteria of indeterminate values can be expected to lead to some changes in the findings, which is also true for the current study. Using a more conservative alpha level decreases the number of values that are statistically different in the traditional null hypothesis test, and as a result reducing the number of variables and sides categorized as substantially different. Excluding values that are statistically indeterminate lead to only one change, namely that Denver was classified as "different" rather than similar. A comprehensive analysis of all possible conditions and changes is beyond the scope of this study. These changes due to different conditions are, however, very important as they can lead to substantial differences in the results of empirical studies. Thus, future research should address how different alpha levels, inclusion and exclusion criteria and equivalence intervals influence study outcomes and inferences and possible policy implications made from these studies.

Table: 5.14. Differences in Outcomes by Using different Alpha Levels and Changes in the Inclusion and Exclusion Criteria

Site	Original			Indeterminate Excluded			Alpha Level .01			Indeterminate Excluded & Alpha Level .01		
	Different	Similar	Unknown	Different	Similar	Unknown	Different	Similar	Unknown	Different	Similar	Unknown
Dallas			X		X				X		X	
Denver		X		X				X			X	
Indianapolis		X			X				X		X	
Miami		X			X			X			X	
New Orleans		X			X			X			X	
Phoenix	X			X				X		X		
Portland	X			X			X			X		
San Antonio		X			X			X			X	
San Jose		X			X			X			X	

CHAPTER 6

Discussion and Conclusion

This Chapter provides a review of the study purpose, the results that emerged from the statistical analysis, and the inferences that can be drawn. This is followed by a discussion of the implications and limitations of the current study. This Chapter concludes by conferring the contributions and possible extensions of this work and opportunities for future research in this area.

MAJOR GOAL AND POSSIBLE OUTCOMES OF THE STUDY

The major goal of the current study is to assess whether the drug estimates for selected drugs are similar or different between DUF and ADAM. It was hypothesized that the drug use information in the two samples might not be substantially different for two main reasons: (1) both the probability sample of ADAM and the non-probability sample of DUF rely on volunteers; and (2) both DUF and ADAM were only able to sample arrestees who were held in the facility long enough, resulting in a sample of more serious offenders and offenders who did not have the financial means to post bail.

The analysis included the following nine sites: Dallas, Denver, Indianapolis, Miami, New Orleans, Phoenix, Portland, San Antonio, and San Jose. These nine sites were chosen because they have the same catchment area for DUF and ADAM and sufficient data for each time period to allow for a meaningful comparison. The variables included in the analysis are the same for all sites. These variables are urine analysis results (positive/negative) for marijuana, cocaine, and opiates; self-reported drug use within the last 72 hours for marijuana, powder cocaine, crack cocaine, heroin; and self-reported drug use for the question whether the arrestee had ever used marijuana, powder cocaine, crack cocaine, or heroin.

For the purpose of examining the research question, the current study employed equivalence analysis. Using equivalence analysis, it was determined that there existed four possible outcomes for the current study. These four possible outcomes were:

1) DUF and ADAM can be said to be equivalent (Eq)
2) DUF and ADAM are different (D)
3) DUF and ADAM are different and equivalent (D&Eq)
4) DUF and ADAM are not different and not equivalent (indeterminate) (ND&NEq)

The interpretation of the results for each site and for the selected drugs was completed in accordance with the three decision criteria laid out in Chapter Four. These three decision criteria were:

1) The findings of *Equivalent* (Eq) and *Different but Equivalent* (D&Eq) will be interpreted as not substantially different or similar.
2) Sites will be classified as "unable to be assessed" if more than one third of the drug use values fall into the category "*Not Statistically Different and Not Statistically Equivalent*" (ND&NEq).
3) Third, sites will be classified as substantially different if 20% or more of the drug use values show substantial differences (D).

ANALYTICAL STRATEGY OF THE CURRENT STUDY

The current study uses a research strategy that is commonly used in medical research: equivalence testing. Equivalence testing has also become more popular in other fields, however, because it is useful to assess the comparability of scales, groups, and other outcomes. The underlying idea is to test whether two outcomes can be said to be equivalent or whether they are substantially different. To reiterate, substantially different does not simply mean that there is a statistically significant difference using the traditional null hypothesis test, but rather that the difference is of practical importance. For this type of analysis, two simultaneous tests are carried out: the traditional null hypothesis test and the equivalence test. A substantial *difference* can only be established if the traditional test is statistically significant and the equivalence test is not statistically significant. The results can be said to be *equivalent* if the equivalence test is statistically significant and the traditional test is not. A result of *different but equivalent* is typically interpreted as a lack of substantial differences and an indication that the two outcomes are comparable (as discussed in Chapter 4). Finally, it is possible that the results of the analysis show that neither the traditional test nor the equivalence test is statistically significant. In this case it cannot be statistically determined whether the results are different or equivalent. No conclusion can be drawn either way (Tryon, 2001).

Equivalence testing is typically applied to test the comparability of the effect of different drugs or treatments (i.e., established drug versus alternative drug). The current study is different from clinical studies in that superiority of a certain treatment cannot be established. That, however, was not the goal of this study. Instead, the main purpose was to determine whether the percentage outcomes for 11 drug use variables are substantially different or whether they are similar.

MAIN FINDINGS

OVERALL FNDINGS FOR ALL SITES

Out of a total of 99 drug use values, 14 (14%) were found to be *substantially different*, 67 (68%) were classified as *not substantially different or similar,* and 18 (18%) values were deemed *indeterminate* because they were neither statistically different nor statistically equivalent. Thus, the overall analysis of all drug use values suggests that there are no substantial differences because less than 20% of the values were categorized as *different*.

The site with the greatest number of values classified as *equivalent* and *different but equivalent* was New Orleans with nine values (82%), followed by Denver, Portland, San Antonio, and San Jose with eight values (73%). Seven values (64%) were classified as *equivalent* and *different but equivalent* in Indianapolis and Miami. Finally, in Phoenix and Dallas, six values fell into either of these categories. None of the sites had less than six values classified as *equivalent* and *different but equivalent*.

Further, the analysis shows that each site has a certain pattern with regard to the distribution of the drug use estimates across the outcome categories. In Dallas and San Antonio, neither of the drug use values was substantially different. In Indianapolis, Miami, and San Jose, one of the 11 drug use values was substantially different, but it was not the same for the three sites. Rather, at each site a different value was categorized as substantially different. In Denver and New Orleans, two of the drug use values were classified as substantially different. Again, both sites had different drug use values that were classified as substantially different. Portland had three drug use values that were substantially different, and Phoenix had four values that were classified as substantially different. None of the sites had more than four values that fell into this category.

With regard to the outcome category *not different and not equivalent*, the analysis demonstrates that only one site that had more than one-third of the drug use values categorized as indeterminate, which was Dallas with five values (45%). In Indianapolis, Miami, and San Antonio, three values (27%) could not be statistically assessed. In San Jose, two values (18%) fell into this category, and in Denver and Phoenix one value (9%) was classified as indeterminate. The remaining sites, New Orleans and Portland, had no values that were indeterminate. Following the examination of the distribution of drug use values overall, the next analysis step was to take a closer look at the specific drugs to determine whether there are patterns for each drug with regard to their outcomes.

OVERALL FINDINGS BY DRUGS

The overall results for each drug (marijuana, cocaine, and opiates) are that none was substantially different between DUF and ADAM. Specifically, all but one of the drug estimates for marijuana fell into the categories *equivalent* or *different and equivalent*. Thus, 26 of the 27 drug use estimates for marijuana use, including urine test results for marijuana, self-reported marijuana use within the past 72 hours, and self-reported marijuana use over the lifetime were classified into the category *no substantial difference*. In accordance with the decision criteria, the drug use estimates more marijuana can be said to show no substantial differences between DUF and ADAM because less than 20% of the drug use estimates were significantly different and not equivalent.

The drug use estimates for cocaine consisted of urine test results for cocaine, self-reported powder cocaine use within the past 72 hours, self-reported crack cocaine use within the past 72 hours, self-reported powder cocaine use over the lifetime, and self-reported crack cocaine use over the lifetime. Altogether, the analysis included 45 drug use estimates for cocaine. Of those 45 values for cocaine, eight values (18%) fell into the category "Statistically Different and Not Statistically Equivalent;" 25 values (56%) were classified as either "Statistically Equivalent and Not Statistically Different" or "Statistically Different but Statistically Equivalent;" and 12 values (27%) were neither statistically different nor statistically equivalent. Less than 20% of the drug estimates demonstrated substantial differences, and as a result the drug use estimates for cocaine can also be said to show no substantial differences between DUF and ADAM. The results for cocaine were, however, not as clear cut as for marijuana. Specifically, the drug use estimates for cocaine demonstrate some differences, that is, one or more of the drug use values for cocaine were classified as substantially different in Denver, Miami, New Orleans, Phoenix, and Portland.

With regard to opiates, the analysis included the following variables: urine test results for opiates, self-reported heroin use within the past 72 hours, and self-reported heroin use over the lifetime resulting in a total of 27 drug use values. Of these 27 values, five values (19%) showed substantial differences. Sixteen values (59%) were not substantially different; and six values (22%) could not be statistically determined. Overall, the conclusion is that the drug use estimates for opiates are not substantially different because less than 20% of the values demonstrated substantial differences. It is also apparent, however, that there are a few differences. Specifically, one or more of the opiate values was categorized as substantially different in Phoenix, Portland, and San Jose.

The question why there are differences for some sites and for cocaine and opiates but not for marijuana cannot be answered with the data currently available. This question is very problematic to assess given the nature of the issues that are being looked at in this study. It is, however, an important question that should be explored in future research. There are some alternative techniques that could be used, such as Monte Carlo simulation and perhaps a Bayesian approach to this analytic problem.

Simulation studies would permit some exploration of different scenario outcomes under alternative models of distribution of drug use patterns – for example, comparing an aggregate theoretical site derived from the current empirical data for both DUF and ADAM as a comparator. A Bayesian paradigm might also fit here. If one thinks of Bayesian approaches as being a technique for the assessment of informational utility it might be that coupling Bayesian analysis, perhaps with a proportional-reduction-of-error objective, would be a useful avenue. These types of alternative analysis methods will be discussed in more detail later in the discussion. After having assessed the overall findings for each drug, the next step in the analysis was to explore the specific findings for each site.

SITE-SPECIFIC FINDINGS

Dallas was the only site that could not be judged either way because more than one third of the drug use values were *not different and not equivalent* and as a result were classified as indeterminate. The remaining eight sites had 27% or less of the variables that fell into this category. Of these eight sites, two demonstrated substantial differences, that is, in Phoenix and Portland more than 20% of the drug use values were *statistically different and not equivalent*. Specifically, in Phoenix four values (36%) demonstrated substantial differences and in Portland three values (27%) were substantially different. At the sites of Denver, Indianapolis, Miami, New Orleans, San Antonio, and San Jose, less than 20% of the drug use values were classified as substantially different. The analysis suggests that the outcome is site specific, that is, each site has a specific pattern of how the specific drug use values are distributed over the outcome categories. In sum, Dallas could not be assessed; Phoenix and Portland were categorized as substantially different; and Denver, Indianapolis, Miami, New Orleans, San Antonio, and San Jose were categorized as not substantially different or similar.

DISCUSSION OF THE FINDINGS

The results of this study suggest that the overall results for all sites combined are not substantially different and that the outcomes for three drug categories are not substantially different. The site-specific analysis implies that two sites are substantially different, five sites are similar, and one site could not be judged either way. The majority of findings demonstrate that the drug use estimates in the DUF and ADAM data for the three major drugs are similar. There are, however, some differences, which will be discussed in more detail now.

As described above, there are some variables that do show substantial differences. The variables that fell into the category *different* most often were "urine test results for cocaine," "urine test results for opiates," "ever used powder cocaine," and "ever used heroin." The variable "urine test results for cocaine" was substantially different for three sites: Miami, Phoenix, and Denver. For all three sites there was a substantial increase in the number of arrestees who tested positive. Additionally, "ever used powder cocaine" was substantially different in New Orleans and Portland. The results are consistent in that the data shows an increase in cocaine use from 1997/98 to 2000/01. The same is true for "ever used crack cocaine," which demonstrated a substantial increase in New Orleans. These findings suggest that cocaine might have increased within that five-year period. Another variable that showed substantial differences at some sites was "urine test results for opiates." Opiate use significantly increased between the DUF and ADAM data in Phoenix, Portland, and San Jose. Also, the variable "ever used heroin" substantially increased in Phoenix and Portland. The question of why there are differences for those three sites cannot be determined with the current data. There is also no other source of data currently available to researchers to explore this question. Future research should further explore data collection and analytic strategies that might allow for a better analysis of this question. This might be very difficult, however, because the DUF and ADAM data is historic data and cannot be altered at this point.

CONSISTENCY OF FINDINGS

Despite the differences apparent in the data the overall findings suggest that the drug estimates of DUF and ADAM are not substantially different. It is noteworthy that this finding supports what a study by NIJ found, that is, that although the charge distribution for arrestees of the DUF sample differed from the charge distribution of arrestees from the UCR, the drug use estimates derived from the DUF data were almost identical to the estimates for the UCR data (NIJ, 1990). The current findings are also consistent with the study from Anchorage, which suggested that the sample of females was representative of the female arrestee population despite the fact that the sample of females was a convenience sample (Myrstol and Langworthy, 2005).

EXPLAINING THE RESULTS

One explanation for the results of this study, namely that DUF and ADAM are not substantially different, might be that both studies used volunteers and suffered from non-response bias. This is not a problem for drug research alone but for research in general. As stated earlier, in the case of ADAM, the interview refusal rate for all sites combined was 17.5%. Additionally, of the individuals that did choose to participate in the interview (82.5%), 15.6% refused to provide a urine sample (NIJ, 2000). The DUF data shows that approximately 10% of the selected arrestees refused to interview. Of the arrestees that agreed to participate in the survey, about 20% refused to provide a urine sample (NIJ, 1995). Non-respondents constitute a problem for a study if they are different from respondents in ways that bias the study outcome. Research suggests that non-respondents share certain characteristics (Sharp and Feldt, 1959; Hill, 1997). Similarly, volunteers also appear to share certain characteristics.

Although researchers recognize non-response bias, it is rarely quantified. A number of studies that do attempt to quantify non-response bias suggest that non-respondents differ from respondents in various ways. A number of researchers have found differential survey responses, meaning that some population types of the sample have significantly higher response rates than other population subgroups. For instance, Sharp and Feldt (1959) found that younger persons are significantly more likely to respond than older persons. Response rates decline with increasing age. Additionally, response rates varied considerably depending on marital status. Widowed persons were the least likely to grant an interview (74%), followed by adults who had never been married (82%). Married adults with children had the highest response rate (93%).

The Oslo Health Study also assessed non-response patterns and found that a number of population sub-groups were significantly underrepresented. The underrepresented population subgroups were males, young persons, single/never married and divorced/separated persons, persons not born in Norway, persons with lower or unknown education level, persons with a low socio-economic status, and persons receiving disability benefits (Sogaard, et al., 2004). Additionally,

Vivienne (2002) assessed differences between respondents and non-respondents in a survey about alcohol consumption and found that abstainers were overrepresented among the non-respondents, biasing the sample towards individuals who drink alcohol.

Hill, et al. (1997) examined non-response bias in a lifestyle survey and found that respondents and non-respondents varied significantly with regard to current smoking, hazardous alcohol consumption, and lack of moderate or vigorous exercise. More specifically, non-respondents were significantly more likely to be current smokers. This finding was also confirmed by Bostram, et al. (1993), who studied smoking behavior in Sweden, and Smith and Nutbeam (1990). Contrary, respondents showed significantly higher hazardous alcohol consumption. This finding should caution researchers against believing that non-respondents are always engaging in more risky behavior and unhealthy lifestyles than respondents.

Non-response is of concern especially if it is associated with the variable of interest (Oberski, 2008). For example, if the variable of interest is the prevalence of illicit drug abuse among school children and the majority of children who are using illicit drugs are either absent from school or refuse to participate in the interview, the researcher might draw the conclusion that illicit drug abuse among school children is very rare. If the children who were absent and who refused would have truthfully reported their drug abuse, the researcher might have come to a different conclusion.

With regard to DUF and ADAM, it is possible that the similar results of drug use estimates can be attributed to the fact that probability samples suffer from some of the same shortcomings as do non-probability samples, that is, the sample consists of volunteers. Arrestees who did not volunteer to participate in the DUF study might also have refused to complete the interview in the ADAM program. To date, there is very little research that assesses this question. To this author's knowledge, there is only one study from Anchorage that looked at demographic differences between respondents and non-respondents in the ADAM program. They found no differences for the male probability sample and the female convenience sample. Specifically, the female convenience sample was just as representative of the population of female arrestees as the male probability sample was of the population of male arrestees (Myrstol and Langworthy, 2005). This supports the results of the current study, which showed that both DUF and ADAM produced drug use estimates that were not substantially different despite the differences in the sampling design. Even though the current study only examined the drug use information it is likely that other information contained in DUF and ADAM is also not substantially different. This question should be explored in future studies.

IMPLICATION OF THE CURRENT STUDY

The main implication of the current study and the two studies by NIJ and Myrstol and Langworthy (2005) is that the general assumption of researchers that a non-probability sample produces estimates that are substantially different from a probability sample is not necessarily correct. These three studies assessed only one program, DUF and ADAM, but the findings are remarkably similar and imply that researchers can use the drug use data from both studies and explore research questions that have not been assessed yet because the DUF data was said to be unreliable. This implication is strengthened by the fact that these three studies used different analytical strategies and sites and still arrived at similar conclusions.

With regard to policy makers this would imply that the DUF and ADAM data would lead to similar conclusions about drug use prevalence and patterns. Thus, if the main concern is the implementation of programs aiming to reduce drug use, the non-probability sample DUF data would very likely be sufficient. Specifically, the DUF data as well as the ADAM data showed that some drugs are more popular than others across sites, and that these differences were apparent both in DUF and in ADAM. It is likely that policy makers would have implemented the same programs regardless of whether they used DUF or ADAM. This is a crucial finding because budget restraints currently inhibit the collection of data from arrestees nationwide. Due to this lack of data and knowledge, necessary programs are not implemented and drug trends go undetected until they show up in the general population or until the problem has become epidemic.

LIMITATIONS OF THE CURRENT STUDY

First, one of the major limitations stems from the data itself, specifically, only nine sites could be examined because the majority of sites either did not have the same catchment area or had too few cases for the ADAM sample. The implication of this limitation is that the current study cannot make inferences about the similarity or dissimilarity of the DUF and ADAM data for all sites. This means that the results and conclusions of this study are limited to the available sites. It is possible that the analyzed sites are dissimilar from the sites excluded from this analysis. It is also possible that the analyzed sites are not different than the sites not included. At this point this question cannot be answered.

Second, for both DUF and ADAM, it is unknown whether the arrestees who volunteered to participate and who were in the facility long enough to be interviewed are representative of the arrestee population overall with regard to their drug use prevalence and patterns. This question is important and should be assessed in future research to determine whether arrestees who are available and volunteer to participate differ in their drug use habits from arrestees who are either not available or refuse participation. As described earlier, programs are implemented on the basis of what is known about drug use and these programs are only effective if the information they are based upon is accurate. Thus, future research should attempt to collect data from non-respondents and compare it to the individuals who participated in the study.

Third, equivalence analysis has certain limitations in itself. First, researchers have to determine the equivalence margin. At this time there are no standard rules that are applied by all researchers using equivalence testing. Rather, it is a discretionary decision and as a result it is possible to manipulate the results of the study. For example, setting a higher equivalence margin will improve the chances of finding equivalent result. The opposite is true also; researchers looking for substantial differences might choose a very small equivalence margin (Gotzsche, 2006). Thus, researchers suggest that the equivalence margin should be determined based on scientific grounds, past research results, and clinical standards (LeHenann, 2006). These recommendations are fairly vague and leave much room for discretion.

For social scientists it is even more difficult to find an appropriate equivalence margin than for biomedical researchers. Equivalence is more difficult to assess when factors are not completely constant. Even though clinical research often assumes that a certain drug produces a constant outcome this might not necessarily be true (Gotzsche, 2006). This is also true for the current study. The constancy of drug use over time cannot be assumed. As described above, some changes are to be expected due to the natural fluctuations in drug use prevalence and patterns. This has implications for the equivalence margin. For the current study, the equivalence margin was based on previous drug use research and what has been established as a substantial difference with regard to changes in drug use among arrestees. For researchers who are using this type of analysis it is important to find an equivalence margin that is appropriate for the research topic.

Fourth, the current study found no substantial differences for the majority of variables for DUF and ADAM. Also, all of the findings appear to be in congruence with data from other drug survey. Still, it is possible that the differences found are due to the sampling method. Due to the limitations of probability samples, this possibility cannot definitely be excluded. Unfortunately, there is no dataset that could be examined against the DUF and ADAM data to determine the reason for these differences. The question about why some sites and drug use values are substantially different relates to another question. How can we determine the error associated with using the DUF and ADAM data combined and separately? The current study used a 20% margin, that is, if less than 20% of the drug use values were substantially different then the data for this site was said to be similar enough to be considered equivalent. Using this margin, two sites (Phoenix and Portland) were categorized as substantially different. Further, depending on the alpha level and the inclusion or exclusion of the indeterminate values the findings changed. These changes were discussed in detail earlier.

These findings suggest that researchers can use the DUF data for their analysis. There is, however, a certain risk or error associated with doing so. The current analysis has shown that overall 14% of the drug use estimates were substantially different. Thus, if researchers would draw one of the tests at random assuming that they are using equivalent data, there would be a .14 probability that the data is not equivalent. This probability of using data that is believed to be equivalent when it is not differs depending on the drug and site. For marijuana, the probability would be 0 because none of the estimates were substantially different. For cocaine, the probability would be .18, and for opiates it would be .19. With regard to the ten sites, the probability would be .36 for Phoenix, .27 for Portland, .18 for Denver and New Orleans, .9 for Indianapolis, San Jose, and Miami, and .0 for Dallas and San Antonio.

FINAL REMARKS

In spite of these limitations, however, the findings demonstrate that the drug use information collected via a non-probability sampling procedure in DUF are not substantially different than the drug use information collected via a probability sampling procedure in ADAM. As a result, this research presents a contribution to researchers, especially drug use researchers using the DUF and ADAM data and researchers examining hard-to-access populations, policy makers, and law enforcement.

The termination of the DUF and ADAM study and the reinstatement in 2006 by the ONDCP in only ten counties across the United States (now called ADAM II) has robbed researchers of the ability to track drug using behaviors among arrestees and inform police agencies and policy makers of changes in drugs used by arrestees and prevalence of drug use across different geographic areas at the local level. The current study hopes to lay the groundwork for the implementation of a national study that systematically tracks drug use patterns among arrestees – similar to the NSDUH and the MTF. Considering the budget crisis, it is understood that this national study must be cost-effective, yet provide valid data. One solution might be to supplement a probability sample with a convenience sample.

In fact, some researchers have suggested that it might be possible to use a convenience sample to supplement a probability sample as a means of saving research money and to produce a sample with a smaller mean squared error than would be possible with the probability sample given cost and time restraints The bias in the convenience sample can be reduced, for example, by using the known population variables to calibrate the convenience sample (Kalton and Kasprzyk, 1986). This approach allows researchers to compute inferential statistics from the probability sample, including the calculation of confidence intervals, standard errors, and the representativeness of data with regard to the population of interest and draw a large enough sample that allows researchers to assess a variety of research questions.

Considering the fact that the federal government spends $50 million to study drug abuse among the general population who use drugs rarely, it is reasonable to expect that the government also examines drug abuse among arrestees, who use drugs at much higher rates than the general population and school children.

Appendix A - Demographic Profile by Site

Table A.1. Dallas, TX

	DUF 97/98		ADAM 00/01	
	N	%	N	%
Race				
Black	903	58.4	394	49.1
White	450	29.1	280	34.9
Hispanic	152	9.8	119	14.8
Other	14	.9	9	1.1
Not obtained	28	1.8		
Employment				
Full time	892	57.7	473	59.0
Part time	269	17.4	81	10.1
Unemployed	71	4.6	189	23.6
Other	305	19.7	58	7.2
Not obtained	197	12.7	1	0.1
High school Graduate				
Yes	904	58.4	576	71.8
Offense Category				
Violent Offense	448	29.0	156	19.5
Property Offense	675	43.6	160	20.0
Drug Offense	468	30.3	201	25.1
Other Offense	234	15.1	283	35.2
Age in years (mean)	30		30	
N	1,547		802	

Table A.2. Denver, CO

	DUF 97/98		ADAM 00/01	
	N	%	N	%
Race				
Black	662	34.7	350	26.5
White	545	28.6	364	27.6
Hispanic	645	33.8	547	41.5
Other	42	2.2	57	4.3
Not obtained	14	0.7	1	0.1
Employment				
Full time	979	51.3	667	50.6
Part time	392	20.5	181	13.7
Unemployed	141	7.4	355	26.9
Other	382	20.0	116	8.8
Not obtained	14	0.7		0.0
High school Graduate				
Yes	1,028	53.9	858	65.0
Offense Category				
Violent Offense	524	27.5	313	23.7
Property Offense	361	18.9	247	18.7
Drug Offense	565	29.6	274	20.8
Other Offense	458	24.0	485	36.8
Age in years (mean)	32		32	
N	1,908		1,319	

Table A.3. Indianapolis, IN

	DUF 97/98		ADAM 00/01	
	N	%	N	%
Race				
Black	884	57.2	771	56.6
White	586	37.9	577	42.4
Hispanic	64	4.1	11	0.8
Other	9	0.6	3	0.2
Not obtained	2	0.1		0.0
Employment				
Full time	846	54.8	771	56.6
Part time	280	18.1	182	13.4
Unemployed	116	7.5	281	20.6
Other	277	17.9	120	8.8
Not obtained	26	1.7		0.0
High school Graduate				
Yes	824	53.3	848	62.3
Offense Category				
Violent Offense	404	26.1	359	26.4
Property Offense	372	24.1	291	21.4
Drug Offense	326	21.1	339	24.9
Other Offense	443	28.7	17	1.2
Age in years (mean)	31		31	
N	1,545		1,362	

Table A.4. Miami, FL

	DUF 97/98		ADAM 00/01	
	N	%	N	%
Race				
Black	588	46.2	283	52.9
White	200	15.7	231	43.2
Hispanic	480	37.7	21	3.9
Other				
Not obtained	4	0.3		
Employment				
Full time	627	49.3	296	55.3
Part time	257	20.2	65	12.1
Unemployed	118	9.3	134	25.0
Other	263	20.7	40	7.5
Not obtained	7	0.6		
High school Graduate				
Yes	512	40.3	359	67.1
Offense Category				
Violent Offense	446	35.1	119	22.2
Property Offense	296	23.3	129	24.1
Drug Offense	404	31.8	123	23.0
Other Offense	126	9.9	164	30.7
Age in years (mean)	33		33	
N	1,272		535	

Table A.5. New Orleans, LA

	DUF 97/98		ADAM 00/01	
	N	%	N	%
Race				
Black	1,704	87.0	1,070	86.6
White	216	11.0	158	12.8
Hispanic	15	0.8	2	0.2
Other	12	0.6	4	0.3
Not obtained	12	0.6	2	0.2
Employment				
Full time	905	46.2	556	45.0
Part time	454	23.2	204	16.5
Unemployed	73	3.7	321	26.0
Other	503	25.7	155	12.5
Not obtained	24	1.2		
High school Graduate				
Yes	847	43.2	670	54.2
Offense Category				
Violent Offense	628	32.1	176	14.2
Property Offense	728	37.2	214	17.3
Drug Offense	187	9.5	256	20.7
Other Offense				
Age in years (mean)	30		30	
N	1,959		1,236	

Table A.6. Phoenix, AZ

	DUF 97/98		ADAM 00/01	
	N	%	N	%
Race				
Black	221	13.7	340	11.9
White	790	49.0	1,585	55.6
Hispanic	520	32.3	730	25.6
Other	77	4.8	188	6.6
Not obtained	3	0.2	7	0.2
Employment				
Full time	953	59.2	1,597	56.0
Part time	210	13.0	313	11.0
Unemployed	106	6.6	674	23.6
Other	339	21.0	264	9.3
Not obtained	3	0.2	2	0.1
High school Graduate				
Yes	956	59.3	1,914	67.2
Offense Category				
Violent Offense	271	16.8	607	21.3
Property Offense	414	25.7	625	21.9
Drug Offense	317	19.7	703	24.7
Other Offense	609	37.8	914	32.1
Age in years (mean)	31		31	
N	1,611		2,850	

Table A.7. Portland, OR

	DUF 97/98		ADAM 00/01	
	N	%	N	%
Race				
Black	421	30.0	318	24.6
White	865	61.7	837	64.6
Hispanic	78	5.6	95	7.3
Other	35	2.5	41	3.2
Not obtained	2	0.1	4	0.3
Employment				
Full time	478	34.1	480	37.1
Part time	312	22.3	152	11.7
Unemployed	80	5.7	479	37.0
Other	527	37.6	184	14.2
Not obtained	4	0.3		
High school Graduate				
Yes	828	59.1	963	74.4
Offense Category				
Violent Offense	217	15.5	269	20.8
Property Offense	225	16.1	219	16.9
Drug Offense	407	29.1	448	34.6
Other Offense			11	0.8
Age in years (mean)	33		31	
N	1,401		1,295	

Table A.8. San Antonio, TX

	DUF 97/98		ADAM 00/01	
	N	%	N	%
Race				
Black	203	11.0	142	12.7
White	606	32.9	396	35.3
Hispanic	995	54.1	577	51.4
Other	6	0.3	4	0.4
Not obtained	25	1.4	3	0.3
Employment				
Full time	1,043	56.7	691	61.6
Part time	306	16.6	119	10.6
Unemployed	216	11.7	207	18.4
Other	272	14.8	103	9.2
Not obtained	3	0.2	2	0.2
High school Graduate				
Yes	874	47.5	685	61.1
Offense Category				
Violent Offense	449	24.4	145	12.9
Property Offense	453	24.6	150	13.4
Drug Offense	281	15.3	198	17.6
Other Offense	657	35.7	627	55.9
Not obtained			2	0.2
Age in years (mean)	29		29	
N	1,840		1,122	

Table A.9. San Jose, CA

	DUF 97/98		ADAM 00/01	
	N	%	N	%
Race				
Black	147	11.1	149	11.9
White	424	32.1	379	30.4
Hispanic	568	43.1	574	46.0
Other	170	12.9	148	11.9
Not obtained	10	0.8	144	11.5
Employment				
Full time	739	56.0	743	59.5
Part time	206	15.6	121	9.7
Unemployed	158	12.0	317	25.4
Other	214	16.2	111	8.9
Not obtained	2	0.2		
High school Graduate				
Yes	657	49.8	969	77.6
Offense Category				
Violent Offense	551	41.8	383	30.7
Property Offense	338	25.6	197	15.8
Drug Offense	168	12.7	427	34.2
Other Offense	338	25.6	239	19.2
Not obtained	8	0.6	2	0.2
Age in years (mean)	31		32	
N	1,840		1,122	

Table B.1. Dallas, TX

Variable	DUF				ADAM			
	1997		1998		2000		2001	
	N	%	N	%	N	%	N	%
Urine Test								
Marijuana	429	43.8	243	42.9	177	33.8	85	30.6
Cocaine	311	31.7	161	28.4	149	28.4	68	24.5
Opiates	42	4.3	12	2.1	21	4.0	13	4.7
Self-Reported Drug Use Within 72 Hours								
Marijuana	321	32.8	184	32.5	121	23.1	61	21.9
Cocaine	62	6.3	31	5.5	34	6.5	13	4.7
Crack	133	13.6	58	10.2	60	11.5	37	13.3
Heroin	24	2.4	9	1.6	12	2.3	4	1.4
PCP	8	0.8	5	0.9	5	1.0	0	0.0
Amphetamines	24	2.4	13	2.3	1	0.2	1	0.4
Barbiturates	6	0.6	3	0.5	1	0.2	0	0.0
Ever Used Drug								
Marijuana	794	81.0	418	73.7	369	70.4	180	64.7
Cocaine	334	34.1	176	31.0	172	32.8	88	31.7
Crack	271	27.7	146	25.7	144	27.5	68	24.5
Heroin	97	9.9	59	10.4	45	8.6	30	10.8
N	980		567		524		278	

Table B.2. Denver, CO

Variable	DUF				ADAM			
	1997		1998		2000		2001	
	N	%	N	%	N	%	N	%
Urine Test								
Marijuana	394	41.4	394	41.2	254	41.1	279	39.8
Cocaine	383	40.2	379	39.6	210	34.0	230	32.8
Opiates	34	3.6	38	4.0	23	3.7	38	5.4
Self-Reported Drug Use Within 72 Hours								
Marijuana	259	27.2	296	31.0	175	28.3	225	32.1
Cocaine	62	6.5	78	8.2	32	5.2	44	6.3
Crack	161	16.9	129	13.5	73	11.8	92	13.1
Heroin	26	2.7	32	3.3	14	2.3	20	2.9
PCP	0	0.0	0	0.0	0	0.0	0	0.0
Amphetamines	12	1.3	12	1.3	5	0.8	6	0.9
Barbiturates	4	0.4	4	0.4	5	0.8	1	0.1
Ever Used Drug								
Marijuana	814	85.5	818	85.6	478	77.3	526	75.0
Cocaine	465	48.8	454	47.5	258	41.7	323	46.1
Crack	382	40.1	366	38.3	218	35.3	252	35.9
Heroin	151	15.9	147	15.4	95	15.4	105	15.0
N	952		956		618		701	

Table B.3. Indianapolis, IN

Variable	DUF				ADAM			
	1997		1998		2000		2001	
	N	%	N	%	N	%	N	%
Urine Test								
Marijuana	411	43.9	273	44.9	302	47.9	361	49.3
Cocaine	295	31.5	207	34.0	187	29.7	236	32.2
Opiates	28	3.0	11	1.8	23	3.7	46	6.3
Self-Reported Drug Use Within 72 Hours								
Marijuana	329	35.1	197	32.4	175	27.8	236	32.2
Cocaine	29	3.1	21	3.5	17	2.7	24	3.3
Crack	111	11.8	69	11.3	54	8.6	77	10.5
Heroin	16	1.7	3	0.5	4	0.6	14	1.9
PCP	1	0.1	1	0.2	1	0.2	1	0.1
Amphetamines	13	1.4	4	0.7	2	0.3	1	0.1
Barbiturates	16	1.7	15	2.5	2	0.3	1	0.1
Ever Used Drug								
Marijuana	747	79.7	491	80.8	507	80.5	597	81.6
Cocaine	325	34.7	212	34.9	198	31.4	235	32.1
Crack	299	31.9	189	31.1	162	25.7	234	32.0
Heroin	119	12.7	45	7.4	39	6.2	51	7.0
N	937		608		630		732	

Table B.4. Miami, FL

Variable	DUF				ADAM	
	1997		1998		2000	
	N	%	N	%	N	%
Urine Test						
Marijuana	270	31.6	122	29.2	189	35.3
Cocaine	388	45.4	198	47.4	240	44.9
Opiates	18	2.1	10	2.4	25	4.7
Self-Reported Drug Use Within 72 Hours						
Marijuana	192	22.5	92	22.0	118	22.1
Cocaine	100	11.7	44	10.5	52	9.7
Crack	124	14.5	65	15.6	65	12.1
Heroin	14	1.6	5	1.2	19	3.6
PCP	1	0.1	0	0.0	1	0.2
Amphetamines	0	0.0	0	0.0	1	0.2
Barbiturates	5	0.6	3	0.7	0	0.0
Ever Used Drug						
Marijuana	538	63.0	267	63.9	322	60.2
Cocaine	329	38.5	159	38.0	198	37.0
Crack	219	25.6	108	25.8	117	21.9
Heroin	70	8.2	27	6.5	48	9.0
N	854		418		535	

Table B.5. New Orleans, LA

Variable	DUF				ADAM			
	1997		1998		2000		2001	
	N	%	N	%	N	%	N	%
Urine Test								
Marijuana	381	38.3	368	38.1	280	46.4	281	44.5
Cocaine	456	45.9	440	45.6	205	33.9	227	35.9
Opiates	106	10.7	126	13.1	92	15.2	95	15.0
Self-Reported Drug Use Within 72 Hours								
Marijuana	318	32.0	331	34.3	194	32.1	238	37.7
Cocaine	98	9.9	68	7.0	26	4.3	45	7.1
Crack	165	16.6	174	18.0	66	10.9	84	13.3
Heroin	84	8.5	97	10.1	67	11.1	71	11.2
PCP	1	0.1	3	0.3	0	0.0	0	0.0
Amphetamines	3	0.3	3	0.3	1	0.2	1	0.2
Barbiturates	7	0.7	10	1.0	1	0.2	1	0.2
Ever Used Drug								
Marijuana	762	76.7	752	77.9	475	78.6	462	73.1
Cocaine	332	33.4	337	34.9	167	27.6	178	28.2
Crack	310	31.2	302	31.3	154	25.5	162	25.6
Heroin	190	19.1	217	22.5	132	21.9	122	19.3
N	994		965		604		632	

Table B.6. Phoenix, AZ

Variable	DUF				ADAM			
	1997		1998		2000		2001	
	N	%	N	%	N	%	N	%
Urine Test								
Marijuana	299	30.4	202	32.2	423	33.4	601	37.9
Cocaine	318	32.3	195	31.1	378	29.9	408	25.8
Opiates	92	9.4	36	5.7	76	6.0	101	6.4
Self-Reported Drug Use Within 72 Hours								
Marijuana	211	21.5	143	22.8	276	21.8	468	29.5
Cocaine	65	6.6	41	6.5	62	4.9	80	5.1
Crack	150	15.3	86	13.7	175	13.8	184	11.6
Heroin	80	8.1	38	6.1	62	4.9	67	4.2
PCP	3	0.3	1	0.2	7	0.6	2	0.1
Amphetamines	17	1.7	18	2.9	22	1.7	19	1.2
Barbiturates	12	1.2	5	0.8	6	0.5	4	0.3
Ever Used Drug								
Marijuana	813	82.7	498	79.3	979	77.3	1,320	83.3
Cocaine	494	50.3	309	49.2	619	48.9	830	52.4
Crack	428	43.5	242	38.5	504	39.8	622	39.3
Heroin	231	23.5	122	19.4	229	18.1	269	17.0
N	983		628		1,266		1,584	

Table B.7. Portland, OR

Variable	DUF				ADAM			
	1997		1998		2000		2001	
	N	%	N	%	N	%	N	%
Urine Test								
Marijuana	246	38.1	280	37.0	212	35.3	246	35.4
Cocaine	238	36.9	220	29.1	130	21.6	176	25.4
Opiates	89	13.8	118	15.6	80	13.3	67	9.7
Self-Reported Drug Use Within 72 Hours								
Marijuana	169	26.2	178	23.5	118	19.6	202	29.1
Cocaine	59	9.1	52	6.9	24	4.0	41	5.9
Crack	104	16.1	92	12.2	39	6.5	89	12.8
Heroin	73	11.3	81	10.7	41	6.8	43	6.2
PCP	0	0.0	2	0.3	1	0.2	0	0.0
Amphetamines	28	4.3	10	1.3	7	1.2	7	1.0
Barbiturates	5	0.8	5	0.7	1	0.2	1	0.1
Ever Used Drug								
Marijuana	562	87.1	663	87.7	514	85.5	575	82.9
Cocaine	348	54.0	410	54.2	278	46.3	324	46.7
Crack	301	46.7	323	42.7	235	39.1	297	42.8
Heroin	184	28.5	251	33.2	150	25.0	164	23.6
N	645		756		601		694	

Table B.8. San Antonio, TX

Variable	DUF				ADAM			
	1997		1998		2000		2001	
	N	%	N	%	N	%	N	%
Urine Test								
Marijuana	319	34.3	374	41.1	177	34.0	232	38.5
Cocaine	244	26.2	245	27.0	124	23.8	160	26.6
Opiates	96	10.3	87	9.6	47	9.0	55	9.1
Self-Reported Drug Use Within 72 Hours								
Marijuana	225	24.2	235	25.9	118	22.7	177	29.4
Cocaine	83	8.9	80	8.8	39	7.5	56	9.3
Crack	30	3.2	22	2.4	15	2.9	22	3.7
Heroin	52	5.6	42	4.6	30	5.8	35	5.8
PCP	1	0.1	0	0.0	1	0.2	1	0.2
Amphetamines	10	1.1	11	1.2	1	0.2	4	0.7
Barbiturates	8	0.9	7	0.8	2	0.4	1	0.2
Ever Used Drug								
Marijuana	588	63.2	587	64.6	336	64.6	449	74.6
Cocaine	316	33.9	279	30.7	158	30.4	241	40.0
Crack	120	12.9	105	11.6	68	13.1	98	16.3
Heroin	128	13.7	122	13.4	64	12.3	87	14.5
N	931		909		520		602	

Table B.9. San Jose, CA

Variable	DUF				ADAM			
	1997		1998		2000		2001	
	N	%	N	%	N	%	N	%
Urine Test								
Marijuana	256	28.9	107	24.7	167	31.7	271	37.5
Cocaine	120	13.6	35	8.1	71	13.5	79	10.9
Opiates	49	5.5	19	4.4	24	4.6	20	2.8
Self-Reported Drug Use Within 72 Hours								
Marijuana	171	19.3	78	18.0	118	22.4	210	29.1
Cocaine	39	4.4	14	3.2	12	2.3	28	3.9
Crack	40	4.5	5	1.2	23	4.4	29	4.0
Heroin	29	3.3	8	1.8	10	1.9	12	1.7
PCP	5	0.6	4	0.9	4	0.8	16	2.2
Amphetamines	36	4.1	9	2.1	5	1.0	13	1.8
Barbiturates	4	0.5	1	0.2	0	0.0	4	0.6
Ever Used Drug								
Marijuana	599	67.7	285	65.7	354	67.3	537	74.4
Cocaine	332	37.5	168	38.7	208	39.5	345	47.8
Crack	198	22.4	75	17.3	131	24.9	213	29.5
Heroin	96	10.8	45	10.4	55	10.5	92	12.7
N	885		434		526		722	

References

Abiona, T. C., Balogun, J. A., Adefuye A. S., & Sloan, P. E. (209). Pre-incarceration HIV risk behaviours of maleand female inmates. *International Journal of PrisonerHealth, 5 (2),* 59-70.

Agresti, A. & Finlay, B. (2007). Statistical methods for social sciences. Prentice Hall.

Allen, E. & Seaman, C. A. (2006). Different, equivalent, or both. *Quality Progress,* July.

Altman, D. G. & Bland, J. M. (1995). Absence of evidence is not evidence of absence. *British Medical Journal, 311,* 485.

Appel, P. W., Hoffman, J. H., Blane, H. T., Frank, B.,Oldak, R., & Burke, M. (2001). Comparison of self-report and hair analysis in detecting cocaine use in a homeless/transient sample. *Journal of Psychoactive Drugs, 33,* 47-55.

Batanero, C. & Diaz, C. (2006). Methodological and didactical controversies around statistical inference. *Journées de Statistique,* 1-10.

Beckett, K., Nyrop, K., & Pfingst, L. (2006). Race, drugs, and policing: Understanding disparities in drug delivery arrests. *Criminology, 44 (1),* 105-137.

Bennett, T., K. Holloway, & D. Farrington (2008). "The statistical association between drug misuse and crime: A meta-analysis." *Aggression and Violent Behavior*, 13, 107-118

Bickel, W. K., & DeGrandpre, R. J. (1996). *Drug policy and human nature: psychological perspectives on the prevention, management, and treatment of drug abuse.* New York: Plenum Press

Brecht, M. L., Anglin, M. D., & Lu, T. H. (2003). *Estimating drug use prevalence among arrestees using ADAM data: An application of a logistic regression synthetic estimation procedure*. Research Report, National Institute of Justice: U.S. Department of Justice.

Brook, J. S., Whiteman, M., Finch, S. J., & Cohen, P. (1996). Young adult drug use and delinquency: childhood antecedents and adolescent mediators. *Journal of the American Academy of Child & Adolescent Psychiatry, 35*, 1584-1592.

Bureau of Justice Statistics (1998). *Substance abuse and treatment of adults on probation,1995*. NCJ 1666611.

Bureau of Justice Assistance (1999). *Integrating drug testing into a pretrial services system: 1999 update* (Publication No. NCJ 176340). Washington, DC: Bureau of Justice Assistance, Office of Justice Programs, U.S. Department of Justice.

Bureau of Justice Statistics (2000). Census of state and federal correctional facilities. Washington, D.C.: U.S. Department of Justice.

Bureau of Justice Statistics (2004a). *Profile of jail inmates, 2002*, Washington, D.C.: U.S. Department of Justice.

Bureau of Justice Statistics (2004b). *Drug use and dependence, state and federal prisoners, 2004*, Washington, D.C.: U.S. Department of Justice.

Burke, C. (2004). *City of San Diego clean syringe exchange program: Final evaluation report*. San Diego Regional Planning Agency.

Catania, J. A., Turner, H., Pierce, R. C., Golden, E., Stocking, C., Binson, D., & Mast, K. (1993). Response bias in surveys of AIDS-related sexual behavior. In D. G. Ostrow & R. C. Kessler (Eds.), *Methodological issues on AIDS behavioral* research. New York: Plenum.

Centers for Disease Control and Prevention (CDC) (2004). *HIV/AIDS surveillance report, 2003*. Atlanta: U.S. Department of Health and Human Services.

City of San Diego (2001). Clean syringe exchange pilot program. San Diego City Council.

Chen, H, Stephens, R. C., Cochran, D. C., & Huff, H. K. (1997). Problems and solutions for estimating the prevalence of drug abuse among arrestees. *Journal of Drug Issues, 27*, 689-701.

Cleophas, T. J., Zwinderman, A. H., Cleophas, T. F., & Cleophas, E. P. (2009). *Statistics applied to clinical trials.* Springer.

Cohen, J. (1992). A power primer. *Psychological Bulletin, 112 (1)*, 155-159.

Cone, E. J. (1997). New developments in biological measures of drug prevalence. In L. Harrison and A. Hughes (Eds.), *The Validity of Self-Reported Drug Use: Improving the Accuracy of Survey Estimates*. Rockville, MD: National Institute of Drug Abuse.

Cook, S. W., & Selltitz, C. (1964). A multiple-indicator approach to attitude measurement. *Psychological Bulletin*, *62*, 36-55.

Corman, H. & H. N. Mocan (2000). A time-series analysis of crime, deterrence, and drug abuse in New York City. *American Economic Review*, *90 (3)*, 584-604.

Crowne, D. & Marlow, D. (1964). *The approval motive.* New York: Wiley.

Cunningham, J.K. & Liu, L.-M., 2003. Impacts of federal ephedrine and pseudoephedrine regulations on methamphetamine-related hospital admissions. *Addiction, 98,* 1229–1237.

Cunningham, J.K., & Liu, L.-M., 2005. Impacts of federal precursor chemical regulations on methamphetamine arrests. *Addiction, 100,* 479–488.

Decker, S. H., Pennell, S. & Caldwell, A. (1997). *Illegal firearms: Access and use by arrestees.* Research in Brief. National Institute of Justice: U.S. Department of Justice.

Dembo, R.,Williams, L.,Wish, E. D., Berry, E., Getreu, A.,Washburn, M., & Schmeidler, J. (1990). Examination of the relationships among drug use, emotional/psychological problems, and crime among youths entering a juvenile detention center. *The International Journal of the Addictions, 25*, 1301–1340.

DeSimone, J. (2005). Needle exchange programs and drug injection users. *Journal of Policy Analysis and Management, 24 (3),* 559-577.

Des Jarlais, D., C. (1998). Commentary: Validity of self-reported data, scientific methods and drug policy. *Drug and Alcohol Dependence 51*, 265–266

Drug Abuse Warning Network (2011). Illicit drug-related emergency visits in Metropolitan Areas of the United States: 2009. Substance Abuse and Mental Health Services Administration, Office of Applied Studies.

Drug Enforcement Agency (2007a). *DEA staffing and budget.* Available online at:http://www.usdoj.gov/dea/agency/staffing.htm.

Drug Enforcement Agency (2007b). *Organized crime drug enforcement task forces (OCDETF)*. Available online at: http://www.usdoj.gov/dea/programs/ocdetf.htm.

Drug Enforcement Agency (2011). Methamphetamine lab incidents, 2004-2010. United States Department of Justice. Available online at: http://www.justice.gov/dea/concern/map_lab_seizures.html.

Drug Policy Alliance (2006). *Proposition 36: Improving lives, delivering results. A review of the first four years of California's Substance Abuse and Crime Prevention Act.* Available online at: http://www.drugpolicy.org/docUploads/Prop36March2006.pdf.

Drug Reform Coordination Network (2005). *Marijuana law enforcement costs more than $7 billion a year – and doesn't work says new report.* Available online at: http://stopthedrugwar.org/chronicle-old/379/report1.shtml.

Edwards, A. L. (1953). The relationship between the judged desirability of a trait and the probability that the trait will be endorsed. *Journal of Applied Psychology, 37*, 90-103.

Elliott, D. S., & Ageton, S.S. (1980). Reconciling race and class differences in self-reported and official estimates of delinquency. *American Sociological Review,* 45(1), 95–110.

Epstein, J., Klinkenberg, W. D., Wiley, D. & McKinley, L. (2001). Insuring sample equivalence across internet and paper-and-pencil assessments. *Computers in Human Behavior, 17 (3),* 339-346.

European Medicines Agency (2000). *Points to consider in switching between superiority and non-inferiority*. Committee for Proprietary Medicinal Products: The European Agency for the Evaluation of Medicinal Products. Available at: http://www.ema.europa.eu/pdfs/human/ewp/048299en.pdf.

Fagan, J. & K. L. Chin (1990). Violence as regulation of and social control in the distribution of crack. In: M. de la Rosa, E. Y. Lambert, and B. Gropper (Eds.), *Drugs and violence: Causes, correlates, and consequences*. NIDA Research Monograph No. 103, Rockville, MD: National Institute on Drug Abuse.

Fendrich, M., Johnson, T. P. Sudman, S., Wislar, J. S., & Spiehler, V. (1999). Validity of drug use reporting in a high-risk community sample: A comparison of cocaine and heroin survey reports with hair tests. *American Journal of Epidemiology, 149*, 945-962.

Feucht, T. E. & Kyle, G. M. (1996). *Methamphetamine use among adult arrestees: Findings from the Drug Use Forecasting (DUF)*

Program. National Institute of Justice. U.S. Department of Justice.

Franco, C. (2007). Methamphetamine: Legislation and issues in the 109th Congress. In: Gerald H. Toolaney. *New Research on Methamphetamine Abuse*. New York: Nova Science Publisher.

French, M. T. & Martin, R. F. (1996). The cost of drug abuse consequences: A summary of research findings. *Journal of Substance Abuse Treatment, 13(6)*, 453-466.

Gfroerer J, Eyerman J, & Chromy J, (Eds.) (2002). *Redesigning an ongoing national household survey: Methodological issues*. DHHS Publication No. SMA 03–3768. Rockville, MD: Substance Abuse and Mental Health Services Administration, Office of Applied Studies.

Goldstein, P. J. (1990). Crack and homicide in New York City: A conceptually based event analysis, *Contemporary Drug Problems*, 4, 651-687.

Goldstein, P. J. (1987). *Drug-related crime analysis and homicide*. A report to the National Institute of Justice, U.S. Department of Justice.

Golub, A. L., & Johnson, B. D. (1997). Crack's decline. Some surprises across U.S. cities. Research in Brief. National Institute of Justice, U.S. Department of Justice.

Golub, A. L., & Johnson, B. D. (2001). *The rise of marijuana as the drug of choice among youthful adult arrestees*. Research in Brief. National Institute of Justice. U.S. Department of Justice.

Golub, A. L., & Johnson, B. D. (2005). The new heroin users among Manhattan arrestees: Variations by race/ethnicity and mode of consumption. *Journal of Psychoactive Drugs*, *37*, 51- 61.

Grogger, J., & Willis, M. (2000). The emergence of crack cocaine and the rise in urban crime rates. *The Review of Economics and Statistics*, *82*, 519-529.

Hammett, T. M., Harmon, P., & Rhodes, W. (2002). The burden of infectious diseases among inmates of and releasees from US correctional facilities, 1997. *American Journal of Public Health*, *92*, 1789–1794.

Harrell, A. V. (1985). Validation of self-report: the research record in self-report methods of estimating drug use and meeting current challenges to validity. In B. E. Rouse, & N. J. Kozol (Eds.),

National Institute on drug abuse research monograph, vol. Washington, DC: US Government Printing Office.

Harris, R. J. (1997). Significance tests have their place. *Psychological Sciences, 8,* 8-11.

Harrison, L. D. (1992). Trends in illicit drug use in the USA; Conflicting results from national surveys. *International Journal of Addictions, 27,* 817-847.

Harrison, L. D. (1995). The validity of self-reported data on drug use. *The Journal of Drug Issues, 25,* 91-111.

Harrison, L. D., Martin, S. S., Enev, T, & Harrington, D.(2007). *Comparing drug testing and self-report of drug use among youths and young adults in the general population.* Rockville, MD: DHHS Publication No. SMA 07-4249.

Hartley, R., Maddan, S. & Spohn, C. (2007). Prosecutorial discretion: An examination of substantial assistance departures in Federal crack-cocaine and powder cocaine cases. *Justice Quarterly, 24,* 382–407.

Hauck, W. W. & Anderson, S. (1986). A proposal for interpreting and reporting negative studies. *Statistics in Medicine, 5,* 203-209.

Harder, V. S., & Chilcoat, H. D. (2007). Cocaine use and educational achievement: Understanding a changing association over the past 2 decades. *American Journal of Public Health, 97 (10),* 1790-1793.

Henkel, D. (2011). Unemployment and substance use: a review of the literature (1990-2010). *Current Drug Abuse Review*, 4(1), 4-27.

Hersen, M., & Gross, A. M. (Eds.). (2008). *Handbook of clinical psychology: Adults and children (2 volumes).* Hoboken, NJ: John Wiley & Sons, Inc.

Herz, D. C. (2000). *Drugs in the heartland: Methamphetamine use in rural Nebraska.* Research in Brief. National Institute of Justice. U.S. Department of Justice.

Hindelang, M.J., Hirschi, T., & Weis, J.G. (1979). Correlates of delinquency: The illusion of discrepancy between self-report and official measures. *American Sociological Review, 44,* 95–1014.

Hindelang, M.J., Hirschi, T., & Weis, J.G. (1981). *Measuring delinquency.* Beverly Hills: Sage Publications.

Hindin, R., McCusker, J., Vickers-Lahti, M., Bigelow, C., Garfield, F., & Lewis, B. (1994).Radioimmunoassay of hair for determination of cocaine, heroin and marijuana exposure:Comparison with self-report. *International Journal of the Addictions, 29*, p. 771-789.

Hora, P., Schma, W. G. & Rosenthal, J. T.A. (1998). Therapeutic jurisprudence and the drug treatment court movement: Revolutionizing the criminal justice system's response to drug abuse and crime," *Notre Dame Law Review*, 101, 74

Hser, Y.-I., Maglione, M., & Boyle, K. (1999). Validity of self-report of drug use among STD patients, ER patients and arrestees. *American Journal of Drug and Alcohol Abuse*, *25*, 81-91.

Huddleston, W. & D. B. Marlowe (2011). *Painting the current picture: A national report card on drug courts and other problem-solving court programs in the United States.* National Drug Court Institute. Bureau of Justice Assistance.

Huizinga, D., & Elliott, D.S. (1986). Reassessing the reliability and validity of self-report delinquent measures. *Journal of Quantitative Criminology*, 2 (4), 293–327.

Hunt, D., Kuck, S., & Truitt, L. (2006). *Methamphetamine use: Lessons learned.* National Institute of Justice. U.S. Department of Justice.

Hyman, H. (1944). Do they tell the truth. *The Public Opinion Quarterly*, *8*, 557-559.

Inciardi, J. A., Martin, S. S., & Butzin, C. A. (2004). Five-year outcomes of therapeutic community treatment of drug-involved offenders after release from prison. *Crime & Delinquency*, *50*, 88-107.

Jenkins, P., Earle-Richardson, G., Tucker Slingerman, D., & May, J. (2002). Time Dependent Memory Decay. *American Journal of Industrial Medicine, 41,* 98-101.

Johnston, L. D. & O'Malley, P. M. (1997). The recanting of earlier reported drug use by young adults. In L. Harrison and A. Hughes (Eds.), *Validity of Self-Reported Drug Use: Improving the Accuracy of Survey Estimates*, Rockville, MD: National Institute of Drug Abuse.

Johnston, L. D., O'Malley, P. M., & Bachman, J. G. (1997). *National survey results on drug use from the Monitoring the Future study, 1975-1995. Volume II: College Students and Young Adults.* Rockville, M.D.: National Institute of Drug Abuse.

Johnston, L. D., O'Malley, P. M., Bachman, J. G., & Schulenberg, J. E. (2007). *Overall, illicit drug use by American teens continues gradual decline in 2007.* University of Michigan News Service:

Ann Arbor, MI. [Online]. Available online: www.monitoringthefuture.org.

Johnston, L. D., O'Malley, P. M., Bachman, J. G., & Schulenberg, J. E. (2008). *Monitoring the Future national results on adolescent drug use: Overview of key findings, 2007* (NIH Publication No. 086418). Bethesda, MD: National Institute on Drug Abuse, 1-70.

Johnson, E.O., & Schultz, L. (2005). Forward telescoping bias in reported age of onset: an example from cigarette smoking. *International Journal of Methods in Psychiatric Research*, 14 (3): 119-129.

Johnston, L. D., O'Malley, P. M., Bachman, J. G., & Schulenberg, J. E. (2011). *Monitoring the Future national survey results on drug use, 1975-2010. Volume I: Secondary school students*. Ann Arbor: Institute for Social Research, The University of Michigan.

Johnston, L. D., O'Malley, P. M., Bachman, J. G., & Schulenberg, J. E. (2007). *Monitoring the Future national survey results on drug use, 1975-2006. Volume I: Secondary school students* (NIH Publication No. 07-6205). Bethesda, MD: National Institute on Drug Abuse.

Justice Policy Institute (2011). *Addicted to courts: How a growing dependence on drug courts impacts people and communities*. Washington, DC.

Kalton, G. & Kasprzyk, D. (1986). The treatment of missing data. *Survey Methodology,12*, 1–16.

Karberg, J. C. & James, D. C. (2005). *Substance dependence, abuse and treatment of jail inmates, 2002*. U.S. Department of Justice Programs. Bureau of Justice Statistics Special Reports.

Kelly, J. A, Stevenson L. Y., Hauth, A. C., Kalichman, S. C., Diaz, Y. E., Brasfield, T. L., Koob, J. J., & Morgan, M. G. (1992). Community AIDS/HIV reduction: The effects of endorsements by popular people in three cities. *American Journal of Public Health, 82,* 1483 – 1489.

Kleiman, M. A. R. (2004). Flying blind on drug control policy. *Issues in Science and Technology*, Summer 2004.

Krantz, J., Ballard, J., & Scher, J. (1997). Comparing the results of laboratory and World-Wide Web samples on the determinants of female attractiveness. *Behavior Research Methods, Instruments, and Computers, 29(2),* 264–269.

Langan, P. A. & Levin. D. J. (2002). *Recidivism of prisoners release in 1994*. Special Report. Bureau of Justice Statistics.

Leff, H. S., Wieman, D. A., McFarland, B. H., Morrissey, J. P., Rothbard, A., Shern, D. L., Wylie, A. M., Boothroyd, R. A., Stroup, T. S., & Allen, I. E. (2005). Assessment of medicaid managed behavioral health care for persons with serious mental illness. *Psychiatric Services, 56,* 1245-1253.

Le Henanff, A., Giraudeau, B., Baron, G., & Ravaud, R. (2006). Quality of reporting of noninferiority and equivalence randomized trials. *JAMA, 295*, 1147 – 1151.

Link, B. G., & Phelan, J. (1995). Social conditions as fundamental causes of disease. *Journal of Health and Social Behavior; Spec No*, 80–94.

Lord, V. B., Friday, P. C., & Brennan, P. K. (2005). The effects of interviewer characteristics on arrestees' responses to drug-related questions. *Applied Psychology in Criminal Justice, 1*(1), 36- 55

Magnusson, D. & Bergman, L. R. (Eds.) (1990). Data Quality in Longitudinal Research. Cambridge: Cambridge University Press.

Mallender, J., Roberts, E., & Seddon, T. (2002). *Evaluation of drug testing in the criminal justice system in three pilot areas.* Home Office Research Findings No. 176, London: Home Office.

Marroun, H. E., Tiemeier, H., Jaddoe, V. W. V., Hofman, A., Verhulst, F. C., van de Brink, W., & Huizink, A. C. (2011). Agreement between maternal cannabis use during pregnancy according to self-report and urinanalysis in a population-based cohort: The generation r study. *European Addiction Research*, 17(1), 37-43.

Martin, S. S., Butzin, C. A., Saum, C. A., & Inciardi, J. A. (1999). Three-year outcomes of therapeutic community treatment for drug-involved offenders in Delaware: From prison to work release to aftercare. *The Prison Journal*, *79*, 291-293.

McBride, D. C., & Swartz, J. (1990). Drugs and violence. In Ralph Weisheit (Ed.), *Drugs, Crime, and the Criminal Justice System.* Cincinnati: Anderson.

Mieczkowski, T., Barzelay, D., Gropper, B., & Wish, E. (1991). Concordance of three measures of cocaine use in an arrestee population: Hair, urine and self-report. *Journal of Psychoactive Drugs,* 23, 241-249.

Mieczkowski, T., & Newel, R. (1997). Patterns of concordance between hair assays and urine analysis for cocaine: Longitudinal analysis of probationers in Pinellas County, Florida. In L. Harrison and A. Hughes (Eds.), *The Validity of Self-Reported Drug*

Use: Improving the Accuracy of Survey Estimates. Rockville, MD: National Institute of Drug Abuse.

Miller, R. N. and Kuhns, B. (2011). Exploring the impact of medical marijuana laws on the validity of self-reported marijuana use among juvenile arrestees over time. *Criminal Justice Policy Review*, published online 10 January 2011, 1-28.

Moses, L. E. (1992). The reasoning of statistical inference. In D. C. Hoaglin & D. S. Moore (Eds.), *Perspectives on contemporary statistics* (pp. 107-122). Washington, DC: Mathematical Association of America.

Mueller, M. (2011). *A community-based approach to planning implementatio of a drug abuse curriculum for elementary schools*. Dissertation, University of Missouri.

Mumola, C. J. and J. C. Karberg, (2006). *Drug use and dependence, state and federal prisoners*, 2004. Bureau of Justice Statistics, U.S. Department of Justice.

Murphy, S. A., Collins, L. M., & A.J. Rush (2007). Customizing treatment to the patient: Adaptive treatment strategies. *Drug and Alcohol Dependence*, 88(2), S1-S72.

Myrstol, B. and Langworthy, R. (2005). *ADAM-Anchorage data: Are they representative?* Working Paper Number 1. Justice Center Working Papers: University of Alaska Anchorage.

National Drug Intelligence Center (2009). *Situation Report: Pseudoephedrine Smurfing Fuels Surge in Large-Scale Methamphetamine Production in California*. Office of Policy and Interagency Affairs, U.S. Department of Justice.

National Criminal Justice Association (1999). The rising methamphetamine crisis: An examination of state responses. *Policy and Practice, 2(1),* 1-12.

National Drug Intelligence Center (2011). *The economic impact of illicit drug use on American Society*. U.S. Department of Justice.

National Institute of Drug Abuse (1990). *Drugs and violence: Causes, correlates, and consequences. Research Monograph Series 103.* Washington D.C.: National Institute of Health.

National Institute of Drug Abuse (2006). *Monitoring the Future National Survey Results on Drug Use, 1975-2007, Volume II: College Students and Adults Ages 19-45, 2007*. Washington D.C.: National Institute of Health.

National Institute of Justice (1990). *Drug use forecasting – DUF estimates of drug use applied to the UCR.* Washington D.C.: National Institute of Justice.

National Institute of Justice (1993). *Identifying and responding to new forms of drug abuse: Lessons learned from "Crack" and "Ice."* Washington D.C.: National Institute of Justice.

National Institute of Justice (1994). *Drug use forecasting – Annual report on adult and juvenile arrestees.* Washington D.C.: National Institute of Justice.

National Institute of Justice (1995). *Drug use forecasting – Annual report on adult and juvenile arrestees.* Washington D.C.: National Institute of Justice.

National Institute of Justice (1997). *Drug use forecasting – Annual report on adult and juvenile arrestees.* Washington D.C.: National Institute of Justice.

National Institute of Justice (1998). *Drug use forecasting in 24 cities in the United States, 1987-1997* Washington, DC: U.S. Dept. of Justice, National Institute of Justice.

National Institute of Justice (1999). *Drug use forecasting – Annual report on adult and juvenile arrestees.* Washington D.C.: National Institute of Justice.

National Institute of Justice (2000). *Applying the New ADAM Method.* Washington, DC: U.S. Department of Justice.

National Institute of Justice (2001). *2001 Arrestee Drug Abuse Monitoring Annual Report.* Washington, DC: U.S. Department of Justice.

National Institute of Justice (2003). *2000 Arrestee Drug Abuse Monitoring Annual Report.* Washington, DC: U.S. Department of Justice.

National Institute of Justice (2004). *Arrestee Drug Abuse Monitoring (ADAM).* Washington, DC: U.S. Department of Justice.

Nurco, D. N. (1985). A discussion of validity. In B. A. Rouse, N. J. Kozel, & L G. Richards (Eds). *Self-Report Methods of Estimating Drug Use*, Rockville, M.D.: National Institute on Drug Abuse.

Oberski, D. (2008). *Self-selection versus non-response bias in the perceptions of mobility surveys. A comparison using multiple imputation.* The Hague: The Netherlands Institute for Social Research.

Office of National Drug Control Policy (1996). *Treatment protocol effectiveness study*. Washington, DC: Executive Office of the President, Office of National Drug Control Policy.

Office of National Drug Control Policy (2001). *The economic costs of drug abuse in the United States 1992-1998.* Washington, DC: Executive Office of the President.

Office of National Drug Control Policy (2002). *National drug control budget. Executive summary. Fiscal Year 2002.* Executive Office of the President.

Office of National Drug Control Policy (2003). *Drug data summary.* NCJ 191351. Executive Office of the President.

Office of National Drug Control Policy (2007). *National drug control strategy. FY 2008 Budget Summary.* NCJ 216432. Executive Office of the President.

Office of National Drug Control Policy (2008). ADAM II. Annual Report. Executive Office of the President.

Office of National Drug Control Policy (2009). *National drug control budget. Executive summary. Fiscal Year 2010.* Executive Office of the President.

Office of National Drug Control Policy (2010). *Fact Sheet: National Youth Anti-Drug Campaign.* Executive Office of the President.

Office of National Drug Control Policy (2011a). *ADAM II: 2010 Annul Report.* Executive Office of the President.

Office of National Drug Control Policy (2011b). *National Drug Control Strategy 2011.* Executive Office of the President

Office of Safe and Drug Free Schools (2006). *Preliminary report to the secretary on the State grants program.* Available online: http://www.ed.gov/about/bdscomm/list/sdfscac/grantrpt1.html

Parry, H. J., & Crossley , H. M. (1950). Validity of responses to survey questions. *The Public Opinion Quarterly, 14*, 61-80.

Pasveer, K., & Ellard, J. (1998). The making of a personality inventory: help from the WWW. *Behavior Research Methods, Instruments, and Computers, 30(2),* 309–313.

Peters, R.J. Jr, Yacoubian, G.S. Jr, Baumler, E.R., Ross, M. W., & Johnson, R. J. (2002). Heroin use among southern arrestees: Regional findings from the Arrestee Drug Abuse Monitoring Program. *Journal of Addictions & Offender Counseling, 22*, 50-60.

Petersilia, J., Greenwood, P. W., & Lavin, M. (1978). *Criminal careers of habitual felons*. Washington, DC: U.S. Government Printing Office.

Pocock, S. (2003). The pros and cons of noninferiority trials. *Fundamental and Clinical Pharmacology, 17,* 483-490.

Polla, M., Stramesi, C., Pichini, S., Palmi, I., Vignali, C., & Dall'Olio, G. (2009). Hair testing is superior to urine to disclose cocaine conception in driver's license granting. *Forensic Science International*, 189, 41-43.

Prendergast, M. L., & Wexler, H. K. (2004). Correctional substance abuse treatment Programs in California: A historical perspective. *The Prison Journal, 84*, 8-35.

Presser, (1990). Can context changes reduce vote overreporting? *Public Opinion Quarterly, 54*, 586-593.

Resignato, A. J. (2000). Violent crime: A function of drug use or drug enforcement. *Applied Economics, 32*, 681-688.

Reuter, P. (2006). What drug policies cost. Estimating government drug policy expenditures. *Addiction, 101*, 315-322.

Richter, L., & Johnson, P. B. (2001). Current methods of assessing substance use: a review of strengths, problems, and developments. *Journal of Drug Issues, 46*, 34–42.

Riley, K. J. (1997). *Crack, powder cocaine, and heroin: Drug purchase and use patterns in six U.S. cities*. National Institute of Justice. U.S. Department of Justice.

Riley, K.J., Lu, N.T., & Taylor, T.G. (2000). Drug screening: A comparison of urinalysis results from two independent laboratories. *Journal of Drug Issues. 30 (1),*173 – 185.

Roberts, J., Mulvey, E. P., Horney, J., Lewis, J., and Arter, M. L. (2005). A test of two methods of recall for violent events. *Journal of Quantitative Criminology, 21,* 175-193.

Rogers, J. L., Howard, K. I., & Vessey, J. T. (1993). Using significance tests to evaluate equivalence between two experimental groups. *Psychological Bulletin, 113,* 553–565.

Roth, J. A. (1994). *Psychoactive substances and violence*. Research in Brief. U.S. Department of Justice: National Institute of Justice.

Rusticus, S. A. and Lovato, C. Y. (2011). Applying tests of equivalence for multiple group comparisons: Demonstration of the confidence interval approach. *Practical Assessment, Research & Evaluation,*

16(7), Available online: http://pareonline.net/getvn.asp?v=16&n=7.

Sharp, H., & Feldt, A. (1959). Some factors in a probability sample survey of a metropolitan community. *American Sociological Review, 24*, 650-661.

Sloan III, J. J., Bodapati, M. R., & Tucker, T. A. (2004).Respondent misreporting of drug use in self-reports. Social desirability and other correlates. *The Journal of Drug Issues, 34*, 269-292.

Smith, C., & Nutbeam, D. (1990). Assessing non-response bias: A case study from 1985 Welsh Heart Health Survey. *Health Education Research, 5*, 381-386.

Stegner, B. L., Bostrom, A. G., & Greenfield, T. K. (1996). Equivalence testing for use in psychosocial and services research: An introduction with examples. *Evaluation and Program Planning, 19 (3),* 193-198.

Stone, A. A., Turkkan, J. S., Bachrach, C. A., Jobe, J. B., Kurtsman, H. S., & Cain, V. S. (2000). *The science of self-report: Implications for research and practice*. Mahwah, NJ: Lawrence Erlbaum.

Substance Abuse and Mental Health Service Administration (2002). *Results from the 2001 National Survey on Drug Use and Health: National Findings.* Office of Applied Studies, Rockville, MD.

Substance Abuse and Mental Health Service Administration (2002a). *Emergency Department Trends From the Drug Abuse Warning Network, Final Estimates 1994-2001.* Office of Applied Studies, Rockville, MD.

Substance Abuse and Mental Health Service Administration (2002b). *Drug Abuse Warning Network: Development of a New Design.* Methodology Report: Office of Applied Studies. U.S. Department of Health and Human Services.

Substance Abuse and Mental Health Service Administration (2003). *Drug and alcohol information systems: The DASIS report.* Office of Applied Studies. U.S. Department of Health and Human Services.

Substance Abuse and Mental Health Service Administration (2004). Meth Abuse Increases in the Midwest. Big Increases Seen on the East coast also. News Release. Available online at: http://alcoholism.about.com/od/meth/a/blsam040822.htm

Substance Abuse and Mental Health Services Administration. (2005). *Results from the 2005 National Survey on Drug Use and Health.* Department of Health and Human Services.

Substance Abuse and Mental Health Services Administration. (2006). *National Survey of Drug Use and Health: Summary of Methodological Studies.* Department of Health and Human Services

Substance Abuse and Mental Health Service Administration (2011). *Results from the 2010 National Survey on Drug Use and Health: National Findings.* Office of Applied Studies, Rockville, MD.

Sutton, J. E., P.E. Bellair, B. R. Kowalski, R. Light, and D. T. Hutcherson (2011). Reliability and validity of prisoner self-reports gathered using the life event calendar method. *Journal of Quantitative Criminology*, 27(2), 151-171.

Tassipoulos, K., Bernstein, J., Heeren, T., Levenson, S., Hingson, R., & Bernstein, E. (2004). Hair testing and self-report of cocaine use by heroin users. *Addiction*, 99, 590-597.

The National Center on Addiction and Substance Abuse (CASA) at Columbia University. (2010a). *CASA analysis of the Bureau of Justice Statistics bulletin: Prison Inmates at midyear 2007* (NCJ Pub. No. 221944).Washington, DC: U.S. Department of Justice, Office of Justice Programs, Bureau of Justice Statistics.

Tryon, W. W. (2001). Evaluating statistical difference, equivalence, and indeterminacy using inferential confidence intervals: An integrated alternative method of conducting null hypothesis statistical tests. *Psychological Methods, 6,* 371–386.

Tryon, W. W. and Lewis, C. (2009). Evaluating Independent Proportions for Statistical Difference, Equivalence, Indeterminacy, and Trivial Difference Using Inferential Confidence Intervals, *Journal of Educational and Behavioral Statistics, 34 (2),* 171-189.

United States General Accounting Office (1993). *Drug use measurement: strength, limitations, and recommendations for improvement.* Report to the Chairman, Committee on Government Operations, House of Representatives. General Accounting Office, Washington DC.

Vignali, C., Stramesi, C., Vccio, M., & Groppi, A. (2011). Hair testing an self-reported cocaine use. *Forensic Science International*, In Press, Available online June 2011.

Webb, V. J., & Delone, M. A. (1996). Drug use among a misdemeanant population. *Crime, Law and Social Change*, 24(3), p. 241-255.

Westlake, W., J. (1981). Bioequivalence testing—A need to rethink (Reader reaction response). *Biometrics, 37,* 591-593.

Wiens, B., L. (2001). Something for nothing in noninferiority/superiority testing: a caution. *Drug Info Journal, 35,* 241–245.

Willis, J. J., Mastrofski, S. D., & Weisburg, D. (2003).*Compstat in practice: An in-depth analysis of three cities*. Police Foundation.

Wish, E., & Gropper, B. (1990). Drug Testing by the Criminal Justice System. In M. Tonry, & J. Wilson, *Drugs and Crime,* (Eds.). *Vol. 13 of Crime and Justice: A Review* of *Research.* Chicago: University of Chicago Press.

Yacoubian, G. S. (2003a). Correlates of benzodiazepine use among a sample of arrestees surveyed through the Arrestee Drug Abuse Monitoring (ADAM) program. *Substance Use & Misuse*, *38*, 127-139.

Yacoubian, G. S. (2003b). Does the calendaring method enhance drug use reporting among Portland arrestees? *Journal of Substance Use, 8 (1),* 27-32.

Yacoubian, G. S. (2004). The sins of ADAM: Toward a new national criminal justice drug use surveillance system. *International Journal of Drug Testing*, *3*, 1-32.

Yacoubian, G. S., Urbach, B. J., Larsen, K. L., Johnson, R. J., & Peters Jr., R. J. (2000). A comparison of drug use between prostitutes and other female arrestees. *Journal of Alcohol and Drug Education*, 46, 12-25.

Yacoubian, G. S., Peters, R. J., Urbach, B. J., & Johnson, R.J. (2002). Comparing drug use between welfare-receiving arrestees and non-welfare-receiving arrestees. *Journal of Drug Education, 32*, 139-147.

Yang, Y. M. (2004). *Survey errors and survey costs: Experience from surveys of arrestees*. American Statistical Association Section on Survey Research Methods, 4656- 4659.

Zerbe, W. J., & Paulhus, D. L. (1987). Socially desirable responding in organization behavior: A reconception. *Academy of Management Review*, *12*, 250-264.

Index

L

M

N

O

P

R

S

T

U

V

W

Y

CPSIA information can be obtained at www.ICGtesting.com
Printed in the USA
BVOW081207270313

316430BV00002B/3/P

9 781593 325190